T0231173

A Physical Approach to
Color Image Understanding

A Physical Approach to Color Image Understanding

Gudrun J. Klinker
Digital Equipment Corporation
Cambridge Research Lab
Cambridge, Massachusetts

CRC Press
Taylor & Francis Group
Boca Raton   London   New York

CRC Press is an imprint of the
Taylor & Francis Group, an **informa** business

AN A K PETERS BOOK

CRC Press
Taylor & Francis Group
6000 Broken Sound Parkway NW, Suite 300
Boca Raton, FL 33487-2742

**Visit the Taylor & Francis Web site at
http://www.taylorandfrancis.com**

**and the CRC Press Web site at
http://www.crcpress.com**

ISBN 13: 978-1-56881-013-3 (hbk)
ISBN 13: 978-0-429-06535-4 (ebk)

DOI: 10.1201/9781439864685

To Georg, Kai, and Jens

Contents

List of Figures   xi

List of Color Figures   xiii

List of Tables   xvii

Preface   xix

1 Introduction   1
   1.1 Intrinsic Physical Models for Computer Vision   2
   1.2 A Physical Approach to Color Image Understanding   5
   1.3 Preview of Results   7
   1.4 Outline   7

2 A Physical Reflection Model   9
   2.1 The Physics of Light Reflection   9
   2.2 The Dichromatic Reflection Model   13
   2.3 Object Shape and Spectral Variation   15
   2.4 Dimensionality of the Measurement Space   29
   2.5 Material Classes 29
   2.6 Summary   33

3   A Sensor Model   **35**
  3.1   Spectral Integration   36
  3.2   Limited Dynamic Range   37
  3.3   Color Balancing and Spectral Linearization   41
  3.4   Chromatic Aberration   43
  3.5   Examples: Color Clusters from Real Color Images   45
  3.6   Summary   48

4   Color Image Segmentation   **51**
  4.1   Color Image Analysis Guided by the Dichromatic Reflection
        Model   52
  4.2   Generating Initial Estimates for Color Clusters   55
  4.3   Generating Linear Hypotheses   59
  4.4   Exploiting Linear Hypotheses   60
  4.5   Generating Planar Hypotheses   63
  4.6   Exploiting Planar Hypotheses   65
  4.7   Accounting for Camera Limitations   68
  4.8   Optical Effects Beyond the Scope   68
  4.9   Summary   70

5   Separating Pixels into Their Reflection Components   **71**
  5.1   Determining Body and Surface Reflection Vectors   72
  5.2   Generating Intrinsic Reflection Images   82
  5.3   Restoring the Colors of Clipped and Bloomed Pixels   85
  5.4   The Use of Intrinsic Reflection Images   86
  5.5   Summary   87

6   Results and Discussion   **89**
  6.1   Further Results   89
  6.2   Comparison with a Traditional Color Segmentation Algorithm   93
  6.3   Control Parameters   94
  6.4   Simplifying Heuristics   98
  6.5   Limitations of the Dichromatic Theory   100
  6.6   Summary   103

7   Summary and Conclusions   **105**
  7.1   Contributions   105
  7.2   Directions of Future Research   109

8   Related Work from 1988 until 1992   **113**
  8.1   Intrinsic Reflection Images   114
  8.2   Color Image Segmentation   118

8.3   Reflection and Camera Models 119
8.4   Analyzing Other Optical Phenomena 122
8.5   Are We There Yet? 125

A Derivation of the 50%-Heuristic   129

B Tables of Illumination Geometries   133

C Illumination Geometry for $d = \infty$ 139

References   141

Index   161

List of Figures

2.1 Light reflection of pigmented dielectric materials. 10

2.2 Photometric angles. 14

2.3 A mixture of the lights reflected from the material surface
and body. 15

2.4 The shape of the spectral cluster for a cylindrical object: (a)
Surface and body reflection from a cylindrical object, (b)
Spectral cluster on the dichromatic plane spanned by the
spectral power distribution of surface and body reflection
$c_s(\lambda)$ and $c_b(\lambda)$. 17

2.5 Assumed illumination geometry. 19

2.6 Brightest matte point that is visible from the camera. . . . 20

2.7 Body reflection components and their quotient q for $d = 2r$,
$\gamma = 30°$, $g_{max} = 120°$. 21

2.8 Illumination geometry for $d = 2r$, $\gamma = 30°$, $g_{max} = 120°$. . . 21

2.9 Starting point q of the highlight line as a function of phase
angle g and distance d. 23

2.10 Difference vectors between matte and highlight pixels on a
skewed T. 24

2.11 Difference vectors on the dichromatic plane. 26

2.12 Orientations of spectral differences on one object, shown on
the Gaussian spectrum sphere. 26

2.13 Orientations of spectral differences on two objects, shown on
     the Gaussian spectrum sphere.   28
2.14 Classification scheme for materials.   30

3.1 Color cluster in a three-dimensional color space.   37
3.2 Color cluster in the color cube, with color clipping.   38
3.3 Spectral responsivity of CCD-cameras and the human eye. .   42
3.4 Chromatic aberration.   44
3.5 Chromatic aberration as a function of the position in the
     scene.   45

4.1 Using an intrinsic model for image understanding.   54
4.2 Interpretation stages.   55
4.3 Color classes: (a) three-dimensional (volumetric); (b) two-
     dimensional (planar); (c) one-dimensional (linear); (d) zero-
     dimensional (pointlike).   57
4.4 Linear hypothesis.   61
4.5 Proximity heuristic to resolve a color conflict between the
     color clusters of two neighboring matte object areas.   62
4.6 Finding the brightest matte pixel.   64
4.7 A planar slice.   66
4.8 Coplanar color clusters.   67

5.1 Fitting a convex polygon to the color cluster.   74
5.2 Classifying the color pixels.   75
5.3 Globally fitted reflection vectors.   76
5.4 Decomposing a color pixel into its constituent body and sur-
     face reflection components.   83

6.1 Cluster shapes for concave objects.   100
6.2 Light reflection from a concave ellipsoid.   101

A.1 Assumed illumination geometry.   130
A.2 Photometric angles at highlight.   131
A.3 Brightest matte point that is visible from the camera. . . .   132

List of Color Figures

1   Scene with eight plastic objects.
2   Region segmentation by Phoenix, scene with eight plastic objects.
3   Segmentation and intrinsic images using Dichromatic Reflection Model, scene with eight plastic objects.
4   The effect of color balancing and gamma correction on color images.
5   Curvature in the color space, caused by gamma-corrected cameras.
6   Spectral linearization with cubic spline functions.
7   Chromatic aberration measurements on a calibration grid.
8   Three plastic cups under yellow light.
9   Color histogram of three plastic cups under yellow light (clipped colors are shown in blue).
10 Orange cup under yellow light.
11 Color histogram of the orange cup under yellow light (bloomed colors are shown in white).
12 Color histogram of the green donut, exhibiting chromatic aberration (side view onto highlight cluster).
13 Color histogram of the green donut (top view onto highlight cluster).

14   Images and color clusters of glossy paper and a ceramic cup under white light.

15   Color histogram of the scene with eight plastic objects.

16   Color cluster classification for initial image areas, scene with eight plastic objects.

17   Initial grouping into approximate image areas, scene with eight plastic objects.

18   Linear segmentation, scene with eight plastic objects.

19   Planar segmentation, scene with eight plastic objects.

20   Final segmentation, scene with eight plastic objects.

21   Interreflection between the yellow and blue donut, magnified portion of the plastic scene.

22   Influence of interreflection between the yellow and blue donut onto the color histogram of the blue donut, plastic scene.

23   Globally fitted color vectors for the yellow donut of the plastic scene.

24   Globally fitted color vectors for the blue donut of the plastic scene.

25   Body reflection image, plastic scene (local). (Clipped pixels and bloomed pixels are shown in black).

26   Surface reflection image, plastic scene (local). (Clipped and bloomed pixels are shown in black)

27   Noise image, plastic scene (local).

28   Body reflection image, plastic scene (local) (restored clipped and bloomed pixels).

29   Surface reflection image, plastic scene (local) (restored clipped and bloomed pixels).

30   Body reflection image, plastic scene (global) (restored clipped and bloomed pixels)

31   Surface reflection image, plastic scene (global) (restored clipped and bloomed pixels)

32   Profiles of the reflection images along row 226, plastic scene (local).

33   Color image segmentation and reflection analysis, scene with three plastic cups under yellow light.

34   Color image segmentation and reflection analysis, scene with three plastic cups under white light.

35   Color image segmentation and reflection analysis, scene with three plastic cups under pink light.

36   The influence of thresholding intensity to exclude dark pixels, three cups under yellow light.

37   The influence of cylinder width, three cups under white light.

38   The influence of the initial window size, three cups under pink light.

39   The influence of the minimal area size for matte linear clusters, three cups under pink light.

List of Tables

4.1 Summary of color cluster interpretations. 58

5.1 Body and surface reflection vectors of the eight plastic objects under white light (global). 78
5.2 Body and surface reflection vectors of the eight plastic objects under white light (local). 78
5.3 Difference vectors and angles between the surface reflection vectors and the independently measured illumination vector of the eight plastic objects under white light. 79

6.1 Normalized body and surface reflection vectors of the three plastic cups under yellow light (local method). 90
6.2 Normalized body and surface reflection vectors of the three plastic cups under white light (local method). 90
6.3 Normalized body and surface reflection vectors of the three plastic cups under pink light (local method). 91
6.4 Normalized body and surface reflection vectors of the three plastic cups under pink light (local method). 91
6.5 Normalized body and surface reflection vectors of the three plastic cups under yellow light (global method). 91

6.6   Normalized body and surface reflection vectors of the three
        plastic cups under white light (global method).   91
6.7   Normalized body and surface reflection vectors of the three
        plastic cups under pink light (global method).   92
6.8   Control parameters of the algorithm.   95

B.1   Body reflection components and quotient q for $d = 1.01r$. .   133
B.2   Body reflection components and quotient q for $d = 1.05r$. .   134
B.3   Body reflection components and quotient q for $d = 1.1r$. . .   134
B.4   Body reflection components and quotient q for $d = 1.2r$. . .   135
B.5   Body reflection components and quotient q for $d = 1.5r$. . .   135
B.6   Body reflection components and quotient q for $d = 2r$. . . .   136
B.7   Body reflection components and quotient q for $d = 5r$. . . .   136
B.8   Body reflection components and quotient q for $d = \infty$. . . .   137

Preface

This monograph is based on my thesis research at Carnegie Mellon University between 1985 and 1988. The work uses an optical reflection model to analyze the effects of shading and highlights on dielectric objects in a scene and to show how the reflection model can be used to guide a color image analysis system in interpreting color changes in images. The resulting algorithm automatically segments color images into object areas, ignoring color changes along highlights and at internal object edges, where the shading changes abruptly. The algorithm also generates two intrinsic reflection images, one showing the scene without highlights, and the other showing just the highlights.

I suggested in 1988 that these results could be used in other areas of computer vision that are disturbed by highlights, such as stereo vision, and I expected the algorithm to be a useful preprocessor for other computer vision algorithms. Furthermore, I expected that this physics-based approach toward color image understanding might lead to more reliable and more useful image understanding methods than those available at that time.

Four years have passed, and physics-based computer vision research has expanded since then into a lively subfield of computer vision with a rapidly growing community of researchers. How far has it gone? Has it fullfilled the expectations? The last chapter of this monograph presents related work that has emerged between 1988 and 1992, and discusses the overall success of the field.

I enjoyed a wonderful atmosphere and working environment in the Computer Science Department and the Robotics Institute at Carnegie Mellon University. My thesis advisors, Steve Shafer and Takeo Kanade, were always very supportive. I thank them for their guidance and encouragement. Both have taken a great deal of interest in my research, always ready to discuss problems and results and to suggest solutions. Steve Shafer was a constant source of ideas. His energy, optimism, and ever-open door helped me to progress through critical phases of the research. The Dichromatic Reflection Model and its potential for splitting color pixels into reflection components were originally his idea. He also provided a lot of help with linking problems of the algorithm to camera characteristics. Takeo Kanade initiated many discussions on alternative approaches and on the power and limitations of the approach taken in this thesis. With his experience in computer vision, he provided perspective on the goals and achievements of this research.

I am also grateful to the other members of my thesis committee: Katsushi Ikeuchi, Ruth Johnston-Feller, H.-H. Nagel, and Jon Webb. Jon Webb was (with Takeo Kanade) my advisor for the first three years of the Ph.D. program. He spent a great amount of time and care in coaching me through all the qualifiers and getting me started in research activities at CMU. H.-H. Nagel, my advisor at the Universitaet Hamburg (Germany), initially piqued my interest in computer vision through his enthusiastic and careful attitude toward science. Ruth Johnston-Feller spent a lot of time describing material properties to me.

My thesis research was conducted using the equipment, software, and support of the VASC-group of the Robotics Institute at Carnegie Mellon. I am very grateful to all the people in that group for their help in using the laboratory and for creating a good working atmosphere. Financial support was provided by the Jet Propulsion Laboratory (California Institute of Technology)[1], by the National Science Foundation[2], and by the Defense Advanced Research Projects Agency (DOD)[3].

Many thanks go to all my office mates at CMU. They were always ready to provide help in getting through the Ph.D. program, fighting with terminals or software, and engaging in discussions on any aspect of life. I am especially thankful to Keith Gremban, Anoop Gupta, Leonard Hamey, Larry Matthies, Victor Milenkovic, and Carol Novak, who never tired of looking at my research results, discussing its features and problems, and suggesting improvements. Keith Gremban and my husband Georg Klinker provided many useful comments on various drafts of my thesis.

[1] NASA contract 957989.
[2] Grant DCR-8419990.
[3] ARPA Order No. 4976, contract F33615-87-C-1499.

Since then, I have moved on to the Cambridge Research Lab of Digital Equipment Corporation where I am now a member of the Visualization Group. I am very grateful to my managers, Ingrid Carlbom and Victor Vyssotsky, who allowed me to spend a significant amount of time on updating this text. At the same time, I would like to thank my publisher, Alice Peters, for being so patient. Several people have reviewed the new chapter which presents research in physics-based vision after my thesis was concluded. I am grateful to Chris Brown, Jill Crisman, Yue Du, Sang Wook Lee, Shree Nayar, Carol Novak, Steve Shafer, and Richard Szeliski for discussing the manuscript with me. Their comments were very helpful.

Last but not least, I would like to thank my family and friends for their support. I am especially grateful to my husband Georg, our two sons Kai and Jens, and to our two families for encouraging and supporting me and for accepting many long evenings and weekends filled with work rather than fun.

1

Introduction

It is the goal of artificial intelligence and robotics to develop automatic systems (robots) that are able to act in and interact with a complex environment and demonstrate a performance that is at least comparable to that of humans. One of the important subtasks in this field is to gather information about the environment dynamically through sensing devices and to interpret this information so that the robot can (re)act accordingly. In this area, computer vision has concentrated on gathering and interpreting visual information. Image understanding involves interpreting two-dimensional images which are projections of three-dimensional scenes. The goal is to extract useful information from the images to plan (re)actions. Such information includes a description of the scene as a collection of shiny and matte surfaces, smooth and rough, interacting with light, shape, and shadow. Unfortunately, light interacts in many complex ways with matter, producing optical effects such as shadow casting, object shading and highlights, and due to the information lost during the projection process, computer vision has not yet been successful at deriving an appropriate description of such surface and illumination properties from an image. The main reason has been a lack of models rich enough to relate pixels and pixel-aggregates to scene characteristics. Furthermore, most image analysis concentrates on interpreting only intensity data in black-and-white images. Such images do not provide enough information to model optical effects. This book presents an approach to computer vision that uses color information to interpret the effects of shading and highlights in a scene. It exploits a physical reflection model which relates shading and highlights to color variation in the image.

1.1 Intrinsic Physical Models for Computer Vision

When we are asked to describe a picture, such as the one shown in Color Figure 1 (see color insert), we generally give a list of objects and their relative positions in the scene. A second, closer look at the image reveals that we have omitted a lot of detail in our description. Objects cast shadows, possibly upon other objects. The brightness on the objects varies, appearing much darker at object parts that are tilted away from the light than where surfaces face the light directly – an effect known as shading. Moreover, some objects have highlights; we may even see a mirror image of one object on another object. These optical effects are caused by various physical processes by which light interacts with matter. They are generally omitted in a human's description of a scene. It takes an artist's trained eye or a computer vision researcher's experience to be aware of them. The human brain performs an enormous, subconscious step of abstraction when it glosses over them. This book claims and demonstrates that it is essential for a successful computer vision program to explicitly perform this step of abstraction, exploiting an intrinsic physical model of the scene when interpreting the image.

Physical processes in the scene have not been emphasized in the traditional line of computer vision research. It has been common practice to divide the image understanding problem into two phases, a low-level segmentation or feature extraction phase and a higher-level reasoning phase in which the image features are related to object features described in object models of the scene [88, 50]. Within this line of research, image segmentation is considered to be a statistical image processing problem with the major concern being to determine statistically significant changes of pixel values under the presence of noise. The implicit assumption is that such significant changes generally correspond to object boundaries in the scene. An entire school of edge detection and region segmentation theories, both for black-and-white images and for color images, exists along this line of thought [155, 71, 122, 20, 138, 146, 147]. In all cases, the emphasis lies on finding all significant changes of pixel values in the image. Since the relationship between image variations and physical processes in the scene is not considered, the generated edge or region images outline not only material boundaries, but also shadows, highlights, and object edges along which the surface orientation changes abruptly. The segmentation is then passed to a higher-level phase which tries to combine regions across highlights or shadows by matching region segments or other extracted features, such as vertices, with parts of models that have been generated for the scene [50].

Recently, research in computer vision has started to pay more attention to the physical processes in the scene occurring when light interacts with objects. A series of *intrinsic models* has been introduced to the vision community, each model describing how one or more particular physical processes of the scene influence the image data. Some intrinsic models describe photometric scene properties, such as how image intensities depend on shading or highlights [66, 209, 76, 74, 58]. Other models concentrate on the geometric aspects, describing how the shapes of shadows depend on the objects in the scene [166, 92], or how different camera positions affect the image in stereo vision [6], motion analysis [190, 29] and optical flow [70]. Geometric modeling also includes work on the appearance of regular texture patterns in the image [91, 90, 75]. The use of these models has been studied, resulting in a set of image understanding algorithms that are often called *shape-from-x* methods, where *x* is a physical process. Each of the algorithms demonstrates that the properties of physical processes can be measured in images and that they can be exploited to derive information about the shape of the objects from the way in which the pixel values change in the image.

The use of intrinsic physical models provides several advantages over the traditional approach to image understanding. First, the complexity of a particular process or of a combination of processes can be understood, if the properties of the processes are analyzed in the scene. Such an analysis can be used to determine how much and what kind of information will be needed to capture the process characteristics adequately in a set of measurements. The result is the design of an *measurement space* in which each dimension represents an independent set of measurements. In this approach toward image formation and image understanding, the researcher decides what kind of data will be needed to solve the problem, instead of using whatever data is conveniently available. Along these lines, Rubin and Richards have discussed the value of color image analysis over the analysis of black-and-white images [157], and Maloney has presented an analysis of how many dimensions a color space needs to be able to model color constancy for a wide selection of naturally occuring materials [117, 118].

The second advantage of intrinsic models is that they can be extended to include more and more physical processes. When a new process is considered, its properties in the scene have to be modeled and then projected into the measurement space. The distinguishing criteria of the new process in the measurement space need to be determined. If such criteria can be found within the given measurement space, the image analysis algorithm can be extended appropriately. If the new process cannot be distinguished from the previous ones in the current measurement space, further sources of independent data, such as that obtained from a different color band or sensor, or from a different sensor position, have to be determined before the algorithm can be extended to account for the new process.

The third advantage concerns sensor modeling and sensor fusion. Every sensor influences the measurements of scene characteristics in a specific way. It also generally has some technical limitations which may lead to specific sources of measurement errors. A robust image analysis algorithm needs to account for such sensor properties. The intrinsic model can be extended by a sensor model in the same way that new physical processes can be considered.

Finally, modeling physical processes in the scene provides the means to evaluate the strengths and limitations of an intrinsic theory and its implementation. Since the assumptions about the processes in the scene are explicitly stated, the kinds of scenes that can be analyzed by the algorithm can be predicted. The complexity involved in extensions can also be estimated.

In comparison, the assumptions and heuristics used in traditional methods are generally related to image properties instead of scene properties. This makes it much harder to predict and evaluate the performance of such algorithms on scenes such as the one shown in Color Figure 1. For the same reason, it is more difficult to extend the algorithms to master problems with a specific physical effect, such as highlights. Furthermore, since traditional methods do not model physical processes, no information exists that could be used to determine the requirements for the dimensionality of the measurement space. Such approaches generally use whatever image data is available. Finally, since traditional methods analyze image properties instead of scene properties, they generally mix considerations about camera limitations with considerations about object properties, typically using a stochastic approach to model changes in the scene under the presence of camera noise. Because the noise model is built into the mathematical formulation of the algorithm, it is much harder for such methods to account for camera problems other than Gaussian noise or to combine information obtained from different sensors.

In conclusion, the use of intrinsic models seems to provide significant advantages over the traditional approach to image understanding. However, a major restriction of current intrinsic models and shape-from-x methods is that they can only be applied to well selected images that exhibit only properties related to a single physical process. Generally, however, many such physical processes occur simultaneously in the scene, and the image is

a complicated result of the superposition of all of them. Barrow and Tenenbaum have suggested splitting an image into a set of *intrinsic images*, each
of which describes the relationship between one physical property of the
scene and the image data, and then to apply the respective shape-from-x
methods separately to the intrinsic images [7]. It is not clear, however,
how the image can be split into these intrinsic components. To achieve
this, intrinsic models have to be combined into a joint analysis of several
physical processes, and a method has to be designed that is able to extract
and separate the influences of all physical scene processes from an image
simultaneously. The work presented in this monograph and previous work
by Shafer [163], Rubin and Richards [157], and Gershon [45, 47] constitute
the first steps in this direction.

1.2 A Physical Approach to Color Image Understanding

This book uses an intrinsic, physical reflection model [162] to interpret the
effects of shading and highlight reflection in color images and to distinguish
them from color changes across material boundaries. The reflection model
is combined with a sensor model which accounts for the limited dynamic
range of cameras, blooming, gamma-correction, and chromatic aberration.
The reflection model states that shading and highlights can be described
in a two-dimensional spectral measurement space. To distinguish such
variation from changes at material boundaries at least a three-dimensional
measurement space is needed. Accordingly, the effects of shading and highlights are difficult to determine in a straightforward way from black-and-
white images in which intensity is the only dimension of measurements.
This text uses color images with three color bands instead.

An analysis of the spectral behavior of object shading and highlight
reflection reveals that the reflected colors of all points on a single object
are a linear combination (i.e., an additive color mixture) of the object
color and the highlight color, and that therefore the colors lie in a plane
in the color space. Furthermore, an analysis of the geometric properties of
highlights and matte shading indicates that the colors form a cluster within
the plane that looks like a skewed T [99].

The combined reflection and sensor model is the basis of an algorithm to
analyze real color images showing objects with highlights. The algorithm
segments color images along material boundaries, ignoring color changes
between matte and highlight areas, as well as shading changes on an object. Looking for characteristic color clusters from local image areas, the
program generates hypotheses that relate object color, shading, highlights
and camera limitations to the shape of the color clusters in the image areas.

In a verification step, the hypotheses are reapplied to the image and every pixel in or near the image areas is tested for fit to the respective hypothesis. The segmentation method alternates between generating hypotheses about the scene from the image data and verifying whether and how the hypotheses fit the image. In this way, the physical model of light reflection drives the segmentation algorithm to identify local and global properties of the scene incrementally, such as object and illumination colors, and to use them in interpreting pixels in the images. The image interpretation process adapts its sensitivity to local scene characteristics and reacts differently to color and intensity changes at different places in the image. The segmentation results are then used to separate color images into two intrinsic images, one showing the scene without highlights, and the other showing only the highlights.

This physical approach to color image understanding has not been the traditional line of research in color image analysis. Previous color analysis systems have separated a color image segmentation step from a subsequent, physics-based post-processing step [157, 45]. Traditional research in color image segmentation has not accounted for the influence of optical effects on object colors and has not been aware of the unique possiblities for distinguishing such effects through color analysis. Traditional color image segmentation algorithms or color edge detectors generally consider object color to be a constant property of an object. They attribute color variation to random camera noise and to material changes [146, 147].

Color Figure 2 (see color insert) shows the results of applying a traditional color segmentation method to Color Figure 1. The figure was obtained by using Phoenix [165], a segmentation program that recursively splits color images into smaller regions until the regions are homogeneous [146]. To decide how to split regions, Phoenix uses a set of user-defined color features, such as red, green, blue, intensity, hue, saturation, etc., each of which is encoded as a separate image band. When a region is considered for further splitting, Phoenix generates for every feature a histogram of all pixels in the region and looks for distinctive valleys in the histograms. Phoenix then splits the region according to the feature with the most prominent valley. If no significant valley can be found in any feature histogram, the region is considered to be homogeneous. Color Figure 2 was generated by running Phoenix on three features: intensity, hue and saturation. The figure shows that some of the highlights. are separated from the surrounding matte object parts while other highlights are integrated with the matte areas. In addition, the matte areas are sometimes split into dark and bright areas, as can be seen on the right half of the green cup, as well as on the dark red (large) donut and on the green donut. This shows that it is not easy to predict how Phoenix will perform on a given image. The reason for the unpredictable behavior of Phoenix is that its heuristic of choosing a feature for its splitting criterion in an area assumes that significant color variation occurs only between objects. The heuristic is not related to the influences of physical processes, such as highlights and shading.

1.3 Preview of Results

Color Figure 3 (see color insert) shows the segmentation and the intrinsic images that have been generated from the image of the plastic scene in Color Figure 1 using the physical approach to color image understanding described above. The results demonstrate that it is possible to use an intrinsic physical model for low-level image understanding. The model can be used to segment an image and split the image into intrinsic matte and highlight components, even if the influences of shading and highlight reflection overlap in the image.

When compared with the traditional segmentation generated by Phoenix in Color Figure 2, Color Figure 3 demonstrates the value of using an intrinsic physical model. The segmentation image in the upper right quarter of Color Figure 3 indicates very well the boundaries of the objects in the scene. All highlights have been recognized as integral parts of the respective objects. Object shading also has been modeled much better than in the Phoenix segmentation: Dark matte areas have not been separated from bright matte areas. Moreover, the intrinsic images in the lower quarters of Color Figure 3 and the generated hypotheses represent important physical information about the scene. Since the hypotheses include an estimate of the illumination color, they provide necessary information for color constancy algorithms which try to discount the influence of the illumination color on the perceived object colors [109, 119, 33, 59]. The intrinsic images can be useful for a variety of algorithms in computer vision that do not account for highlights in images, such as stereo vision and motion analysis [29, 182]. The images can also be used as a preprocessor for methods to determine object shape from shading or highlights [66, 58]. When combined with such methods to interpret the intrinsic images, this research may lead to physics-based image understanding methods that are both more reliable and more useful than traditional methods.

1.4 Outline

This monograph presents a method to analyze the effects of shading and highlight reflection in color images simultaneously. Chapter 2 describes the

underlying Dichromatic Reflection Model and the geometric analysis of the
physical processes. Chapter 3 extends the model to include the properties
and limitations of cameras in the image formation process. Chapters 4 and
5 describe how the Dichromatic Reflection Model can be used in an image
analysis system. Chapter 4 presents the resulting segmentation method
while Chapter 5 describes how the reflection model provides the framework
to separate color images into two intrinsic images, one showing the scene
without highlights and the other showing only the highlights. Chapter 6
shows and discusses the results of using this approach in analyzing a series
of color images. Chapter 7 summarizes the work presented in the previous
chapters. It presents conclusions and directions of future work. Chapter
8 presents newer related work that emerged between 1988 and 1992, i.e,
after the research described in this text was concluded.

2

A Physical Reflection Model

On its path from a light source to the camera, a light ray is altered in many characteristic ways by the objects in the scene. The camera then encodes the measured light into a color pixel. The goal of image understanding is to use properties and relationships between pixels to interpret the image. For such methods to be successful, it is essential that the reflection processes in the scene be understood and modeled, as well as the sensing characteristics of the camera. This chapter describes the processes by which light interacts with objects in the scene.

2.1 The Physics of Light Reflection from Dielectric, Non-Uniform, Opaque Materials

The work in this book is based on the reflective behavior of dielectric, non-uniform materials. When light interacts with dielectric materials, two kinds of reflection exist: The first takes place when light encounters a change in the refractive index at the *material surface*; the second occurs within a *material body* when light is scattered and selectively absorbed [73, 87, 196, 68]. These two kinds of reflection differ in their geometric and photometric properties. As will be shown, those differences can be exploited for computer vision.

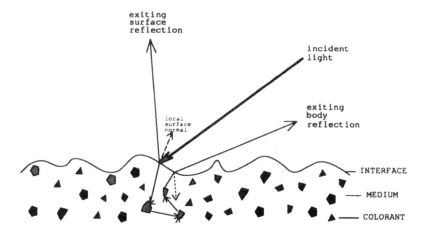

Figure 2.1. Light reflection of pigmented dielectric materials.

2.1.1   Properties of Dielectric, Non-Uniform, Opaque Materials

The most characteristic property of dielectrics is that they do not conduct electricity. Many such materials, e.g., paints, plastics, paper and ceramics, consist of a medium and some embedded pigments, as shown in Figure 2.1 (see color insert). In such *pigmented dielectrics*, the medium comprises the bulk of the matter and is generally approximately transparent, while the pigments, selectively absorb light and scatter it by reflection and refraction. Textiles are also dielectrics, yet they are not pigmented. They are textile fibers to which a dye has been applied. For such *dyed dielectrics*, the entering light is selectively absorbed in the medium in which the dye is dissolved and scattered by the fibers. The basic reflection, scattering and absorption behavior of both pigmented and dyed dielectrics describe the characteristic characteristic appearance of the materials. The reflection model presented in this chapter describes these properties. Although it has been tested so far only for pigmented dielectrics, it ought to be applicable as well to dyed and other dielectrics.

When we look at objects that are made out of such materials, we usually see the reflected light as composed of two distinct colors that typify the highlight areas and the matte object parts. This is a characteristic property of many non-uniform, opaque dielectrics, and the reflection model used in this work capitalizes on this characteristic color change between matte and highlight areas. The next subsections describe the surface and body reflection properties of such material in more detail.

2.1.2 Reflection at the Material Surface

When light hits the surface of a dielectric material, it must first pass
through the interface between the surrounding medium (e.g., air) and the
material. Since the refractive index of the material is generally different
from that of the surrounding medium, some percentage of the incident
light is reflected at the surface of the material. The Dichromatic Reflection Model of Section 2.2 refers to this light reflection process as *surface
reflection*. This process has been called "interface reflection", or "Fresnel
reflection" [163, 59, 97, 98]. Surface reflection is similar, but not identical,
to the concept of "specular reflection" [148, 73], as will be discussed in
Section 2.1.4.

If the interface between the surrounding medium and the material is
completely smooth, the light is reflected from this interface in only one
direction, such that the incident light beam, the exiting light beam and
the surface normal are coplanar and the surface normal bisects the angle
between incident and exiting light. This is the direction of perfect mirror
reflection. However, the interfaces of many materials have some roughness.
The direction of the reflected light depends then on the orientation of the
local surface normals, which vary along the object. In such cases, the
light is scattered to some degree around the global angle of perfect mirror
reflection. Several models have been developed in the physics and computer
graphics communities to describe the geometric properties of light reflection
from rough surfaces [8, 188, 152, 25]. Issues in modeling this process involve
the roughness scale of the surface, compared to the wavelengths of the
incident light, and self-shadowing effects on the surface, which depend on
the viewing direction and the direction of the incident light.

The optical properties of the material also determine the amount and
color of the light that is reflected at the material surface. Fresnel's law
describes how the reflected light depends on the refractive indices of the
material and the surrounding medium, on the incidence angle and on the
polarization of the light [126]. In principle, the color the light reflected
from a dielectric surface is different from the illumination color since the
refractive index is a function of wavelength and thus varies over the light
spectrum [139, 163, 197]. Furthermore, pigment particles may be protruding from the medium at some places [82], so that light incident at such
points changes color according to the absorption properties of the pigment.
Thus, the reflected light may also be a function of the position on the object. However, since the refractive index of most media changes very little
over the visible spectrum, it is common to approximate the refractive index
of the medium as a constant over the visible spectrum. It is also generally
assumed that all pigments are completely embedded in the medium. Under

these assumptions, the light reflected at the surface has the same color as
the illuminating light.

2.1.3   Reflection in the Material Body

For dielectric materials, not all incident light is reflected at the material
surface. Some percentage of the light penetrates into the material body.
In the case of non-uniform dielectrics, the refracted light is scattered and
selectively absorbed in the material body [84, 73, 87]. It travels through
the medium, hitting pigments, fibers or other particles from time to time.
When the light hits a particle, it is scattered. The light keeps hitting
particles and is increasingly scattered until some of it arrives back at the
material surface, where it is partially refracted and reflected in a surface
reflection process. Some fraction of the light then exits from the material
while the rest is reflected back into the material body. This scattering
process in the material body is coupled with selective absorption that either
takes place in the pigments or, in the case of textiles, in the dyed medium.
While the light travels through the medium and bounces back and forth
between the particles, it is increasingly absorbed at wavelengths that are
characteristic for the material.

The Dichromatic Reflection Model refers to the entire process described
above as *body reflection*. Body reflection is similar but not identical to
the concept of "diffuse reflection" [148, 73], as will be discussed in Section 2.1.4. The geometric and photometric properties of body reflection
depend on many factors: the transmitting properties of the medium, the
scattering and absorption properties of the pigments, and the shape and
distribution (including density and orientation) of the pigments [82]. If the
pigments are distributed randomly in the material body, the light exits in
random directions from the body. In the extreme, when the exiting light
is uniformly distributed, the distribution can be described by Lambert's
law. The distribution of the pigments also influences the amount and the
color of the reflected light. If the pigments are distributed randomly in the
material body, the same amount and color will be absorbed on the average
everywhere in the material before the light exits. In such a case, the light
that is reflected from the material body has the same color over the entire
surface.

2.1.4   Terminology

In previous literature on computer vision and computer graphics, the
terms "specular" (or "glossy") and "diffuse" reflection have frequently been
used to refer to surface and body reflection respectively [25, 68, 47]. However, this is not strictly correct. "Specular reflection" refers to light reflec-

tion in the direction of ideal mirror-like reflection, while "diffuse reflection" implies reflection with scattering [148, 73]. This is a purely geometric distinction, while the terms "surface" and "body" reflection refer to different physical processes with distinct spectra. For a very smooth surface, the surface reflection will be specular and the body reflection will be diffuse. However, for typical objects with rough surfaces, the surface reflection will be diffused around the direction of perfect specular reflection, as modeled in [188]. Observing this distinction makes it possible to describe many types of material succinctly: For example, metals, have only surface reflection, which may be specular or diffuse depending on the roughness.

2.2   The Dichromatic Reflection Model

This section presents a mathematical model [163], based on the above discussion that approximates the reflection processes of dielectric, non-uniform, opaque materials. In order for this model to be a suitable basis for work in color image understanding, it makes several restricting and simplifying assumptions on the properties of surface and body reflection, as well as on illumination conditions. Extensions to the model relaxing these assumptions are a subject for future work.

The model assumes that pigments are distributed randomly in the material body and that they are completely embedded in the medium. The surface then exhibits a single spectrum of body reflection and a single spectrum of surface reflection. The model also restricts the illumination conditions of the scene, allowing only one light source and no ambient light or interreflection between objects. These assumptions, in concept, restrict the applicability of the model. However, Section 3.5 shows that the model still provides a reasonable and very useful approximation to the physics of light reflection for many dielectric materials.

Under these assumptions, the light, L, which is reflected from an object point can be described as a mixture of the light L_s reflected at the material surface and the light L_b reflected from the material body.

$$L(\lambda, i, e, g) = L_s(\lambda, i, e, g) + L_b(\lambda, i, e, g). \qquad (2.1)$$

The reflection geometry is illustrated in Figure 2.2. The figure defines several photometric angles [66, 161]: the *incidence angle*, i, between the illumination direction, $\mathbf{I}$, and the surface normal, $\mathbf{N}$, the *exit angle*, e, between $\mathbf{N}$ and the viewing direction, $\mathbf{V}$, and the *phase angle*, g, between $\mathbf{I}$ and $\mathbf{V}$. The parameter λ represents wavelengths of the light spectrum.

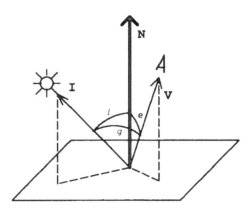

Figure 2.2. Photometric angles.

Since the surface reflection spectrum of dielectric materials depends very little on the illumination geometry [139, 163, 197], the spectral reflection properties of L_s can be approximately separated from its geometric reflection properties. The reflection model thus describes it as a product of a spectral power distribution, $c_s(\lambda)$, and a geometric scale factor, $m_s(i, e, g)$, which describes the intensity of the reflected light. Similarly, the body reflection component L_b can be separated into a spectral power distribution, $c_b(\lambda)$, and a geometric scale factor, $m_b(i, e, g)$. Substituting these terms into equation 2.1 yields the Dichromatic Reflection Model equation:

$$L(\lambda, i, e, g) \;=\; m_s(i, e, g) c_s(\lambda) + m_b(i, e, g) c_b(\lambda). \qquad (2.2)$$

Equation 2.2 describes the light that is reflected from an object point as an additive mixture of two distinct spectral power distributions, $c_s(\lambda)$ and $c_b(\lambda)$, each of which is scaled according to the geometric reflection properties of surface and body reflection. This is shown in Figure 2.3. In the infinite-dimensional vector space of spectral power distributions (each wavelength defines an independent dimension in this vector space [160, 67]), the reflected light can thus be described as a linear combination of the two spectral vectors $c_s(\lambda)$ and $c_b(\lambda)$.

Note that $c_s(\lambda)$ and $c_b(\lambda)$ are not necessarily unit vectors. The vector length depends on the magnitude of the spectral power distribution of the reflected light. For example, white and grey lights define two spectral vectors that have the same orientation but different lengths. The length of the body reflection vector of an object is defined by the magnitude of the

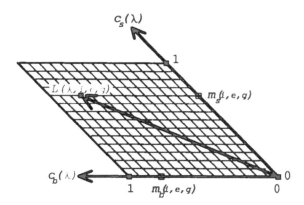

Figure 2.3. A mixture of the lights reflected from the material surface and body.

body reflection component that is reflected from a surface patch pointing in the direction of the light source. Similarly, the length of the surface reflection vector of the object is defined by the maximally existing surface reflection component which is found at surface patches bisecting the phase angle. The geometric scale factors m_s and m_b then vary in the range $[0,1]$, describing what percentage of the maximally reflected light is reflected for a particular set of photometric angles. In the following, the notation $c(\lambda)$ will refer to reflection vectors of arbitrary lengths. Normalized vectors of unit length will be denoted $\bar{c}(\lambda)$.

Many reflection models that have been developed in the physics and computer graphics communities [188, 152, 66, 25] are special cases of the model described here [163]. They replace the geometric variables m_s and m_b by specific functions that approximate the measured reflection data of a chosen set of typical materials. The Dichromatic Reflection Model concentrates on the spectral variables in equation 2.2, $c_s(\lambda)$ and $c_b(\lambda)$, exploiting the spectral difference between them. The geometric factors are not further specified.

2.3   Object Shape and Spectral Variation

The Dichromatic Reflection Model describes the spectral properties separately for every single point on an object. In itself, this description is not yet very helpful in computer vision since it describes the light at each object point by a set of four (so far unknown) factors. It does not provide a mechanism to determine these factors uniquely from the reflected light beam. This section explores the observation that the light reflected from any point on an object is described by the same two spectral vectors and that these two factors in the Dichromatic Reflection Model are thus constant over an object. We show how object shape determines spectral variation on the object and how this spectral variation is related to the body and surface reflection vectors.

2.3.1 Basic Principles

The Dichromatic Reflection Model states that the light spectra of body reflection, $c_b(\lambda)$, and surface reflection, $c_s(\lambda)$, are constant over an entire object, while the geometric scale factors, $m_b(i, e, g)$ and $m_s(i, e, g)$, change with the viewing and illumination angles. Accordingly, $c_b(\lambda)$ and $c_s(\lambda)$ span a *dichromatic plane* in the infinite-dimensional vector space of spectral power distributions, and the light $L(\lambda, i, e, g)$ reflected from any object point lies in this plane. Since $m_s(i, e, g)$ and $m_b(i, e, g)$ are variables in the range from 0 to 1, all light mixtures from an object can be circumscribed in the dichromatic plane by a parallelogram, as shown in Figure 2.3.

An investigation of the geometric properties of surface and body reflection reveals that the light mixtures form a dense spectral cluster in the dichromatic plane. The shape of this cluster is closely related to the shape of the object. This relationship can be used to determine characteristic features of the spectral clusters. The following discussion of spectral histograms uses perspective viewing and illumination geometry. For purposes of illustration, it assumes that body reflection is approximately Lambertian and that surface reflection is describable by a function with a sharp peak around the angle of perfect mirror reflection. Note, however, that this analysis is not limited to a particular geometric reflection model.

Figure 2.4 shows a sketch of a shiny cylinder. The left part of the figure displays the magnitudes of the body and surface reflection components. The curves show the loci of constant body or surface reflection. The darker curves are the loci of constant surface reflection. Since $m_s(i, e, g)$ decreases sharply around the object point with maximal surface reflection, m_{smax}, these curves are shown only in a small object area. The points in this area are called *highlight pixels*. The remaining object points are *matte pixels*. The right part of the figure shows the corresponding spectral histogram in the dichromatic plane. As will be described below, the object points form two linear clusters in the histogram.

Light reflection at matte pixel is primarily determined by the body reflection process. Depending on the surface roughness, matte pixels may contain some amount of surface reflection as the result of light diffusion at the material surface. The following analysis assumes that the surface

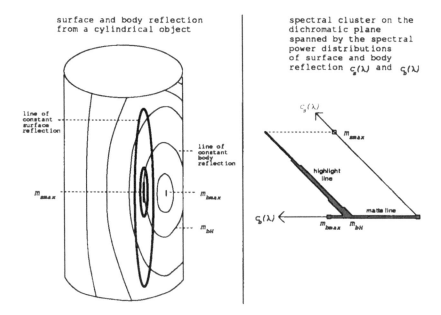

Figure 2.4. The shape of the spectral cluster for a cylindrical object: (a) Surface and body reflection from a cylindrical object, (b) Spectral cluster on the dichromatic plane spanned by the spectral power distribution of surface and body reflection $c_s(\lambda)$ and $c_b(\lambda)$.

reflection component at matte pixels is small and constant. The observed light at these pixels then depends mainly on $c_b(\lambda)$, scaled by $m_b(i, e, g)$ according to the geometric relationship between the local surface normal of the object and the viewing and illumination directions. Consequently, the matte pixels form a *matte line* in the dichromatic plane in the direction of the body reflection vector, $c_b(\lambda)$, as shown in the right part of Figure 2.4.

Highlight pixels exhibit both body reflection and surface reflection. However, since $m_b(i, e, g)$ is much less sensitive to a small change in photometric angles than $m_s(i, e, g)$, the body reflection component is generally approximately constant in a highlight area, as displayed by the curve with label m_{bH} in Figure 2.4. Accordingly, the second term of the Dichromatic Reflection Model equation 2.2 has approximately a constant value, $m_{bH}c_b(\lambda)$, and spectral variation within the highlight comes primarily from varying amounts of $m_s(i, e, g)$. The highlight pixels thus form a *highlight line* in the dichromatic plane in the direction of the surface reflection vector, $c_s(\lambda)$. The line departs from the matte line at position $m_{bH}c_b(\lambda)$, as shown in Figure 2.4. More precisely, the highlight cluster looks like a slim, skewed wedge because of the small variation of the body reflection component over the highlight. When the matte and highlight pixels are combined into a single spectral cluster, this cluster looks like a skewed T. The skewing angle of the T depends on the spectral difference between the body and surface reflection vectors while the position of the highlight line depends on the illumination geometry.

2.3.2   Relationship between Illumination Geometry and Spectral Histogram Shape

There exists a close relationship between the amounts of body and surface reflection on an object and the illumination geometry. This relationship influences the shape of the clusters in the spectral histogram, and it describes constraints that an image analysis program can use when it analyzes spectral variation. Such a constraint, the *50%-heuristic*, will now be derived for the case of spherical objects. It states that, under reasonable assumptions, the highlight cluster starts in the upper 50% of the matte line. It will be used in Chapter 4 for image segmentation to distinguish spectral changes between matte and highlight pixels on one object from spectral changes across material boundaries where matte object areas from different objects meet.

Under the assumption that body reflection can be described by Lambert's cosine law, the body reflection component m_b depends only on the illumination angle. The amount of surface reflection m_s, on the other hand, depends on the positions of both the light source and the camera. Accordingly, different camera positions cause the highlight to move, and thus the surface reflection component to change, while the body reflection component is constant. This shows that the highlight can overlie various amounts of body reflection, depending on the phase angle g between the illumination and viewing direction. If g is very small, i.e., the camera is very close to the light source, the incidence direction of the light at the highlight is close to the surface normal and the underlying amount of body reflection is very high. The highlight line then starts near the tip of the matte line, and the skewed T becomes a skewed L or a "dog leg" [45, 47]. When the phase angle g increases, the amount of underlying body reflection decreases, and the highlight line moves away from the tip of the matte line, exhibiting the T-shape more distinctly.

The following analysis investigates this relationship more precisely for the case of a spherical object with radius r, viewed and illuminated under

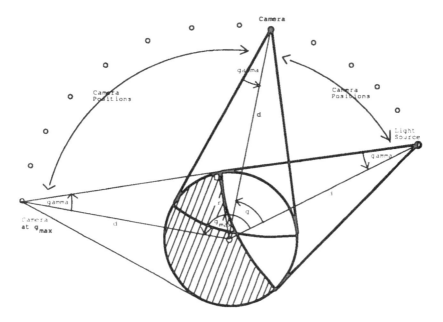

Figure 2.5. Assumed illumination geometry.

perspective projection, as displayed in Figure 2.5. The analysis assumes
that the sphere is illuminated by a point light source at some distance d
from the object center. It further assumes that the camera is positioned
at the same distance d from the object center and that an arbitrary phase
angle g exists between the viewing and the illumination direction. Although
these assumptions on the illumination geometry and on the object shape are
limiting, the following analysis provides general insight into the relationship
between object shapes and the shapes of spectral clusters, illustrating a
prototypical case.

Under the assumed illumination geometry and for Lambertian body reflection, the body reflection component, m_{bH}, under the highlight can be
described as a function of d and phase angle g:

$$m_{bH} = \sqrt{1 - \frac{d^2 \sin^2(g/2)}{d^2 + r^2 - 2dr\cos(g/2)}} \qquad (2.3)$$

$$g \in [-g_{max}, g_{max}] \qquad (2.4)$$

$$g_{max} = 2\arctan\frac{\sqrt{d^2 - r^2}}{r}. \qquad (2.5)$$

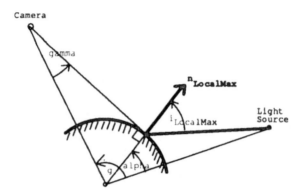

Figure 2.6. Brightest matte point that is visible from the camera.

The start of the highlight line can be related to the length of the matte line which is determined by the brightest matte point on the sphere that is visible from the camera. The body reflection component, $m_{bLocalMax}$, at the brightest visible matte point is given as

$$m_{bLocalMax} = \sqrt{1 - \frac{d^2 \sin^2 \alpha}{d^2 + r^2 - 2dr \cos \alpha}}. \qquad (2.6)$$

with $\alpha = \max(0, g - \frac{g_{max}}{2})$, as shown in Figure 2.6. The ratio, q, of m_{bH} to $m_{bLocalMax}$ describes the starting position of the highlight line, relative to the length of the matte line:

$$q = \frac{m_{bH}}{m_{bLocalMax}}. \qquad (2.7)$$

The details of this derivation are shown in Appendix A.

Figure 2.7 displays the body reflection components and their quotient q as a function of the phase angle g for a fixed distance $d = 2.0r$. This illumination geometry is shown in Figure 2.8. The corresponding table can be found in Appendix B. For $d = 2.0r$, the illuminating cone has a wide angle γ of 30°. Only a small area of the sphere is illuminated. This confines $\|g\|$ to the range of phase angles: $[0, g_{max} = 120°]$. For phase angles $\|g\|$ in the interval $[0, g_{max}/2 = 60°]$, the curve of $m_{bLocalMax}$ is a straight, horizontal line since the globally maximal body reflection component is visible. The body reflection component at the highlight, m_{bH}, on the other hand, decreases rapidly in this range of phase angles. Since $m_{bLocalMax} = 1$, the quotient q has the same value as m_{bH}. The second half of the graph

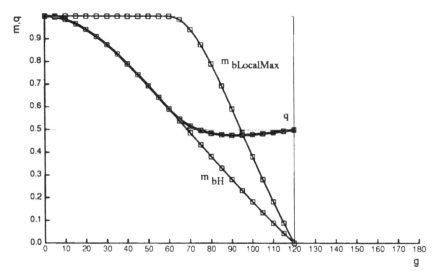

Figure 2.7. Body reflection components and their quotient q for $d = 2r$, $\gamma = 30°$, $g_{max} = 120°$.

for $\|g\| \in [60°, 120°]$ represents illumination geometries under which the brightest matte point is not visible from the camera. The brightest visible body reflection, $m_{bLocalMax}$, decreases in this interval much faster than

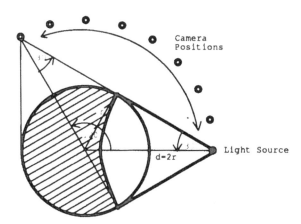

Figure 2.8. Illumination geometry for $d = 2r$, $\gamma = 30°$, $g_{max} = 120°$.

the body reflection under the highlight, m_{bH}. As a consequence, q has an inflection point at $\|g\| = g_{max}/2 = 60°$, decreasing at a slower rate in the second half of the graph. In fact, Figure 2.7 shows that q reaches a minimum and then starts increasing. At the minimum, m_{bH} comprises about 47% of $m_{bLocalMax}$. Accordingly, no phase angle $\|g\|$ for $d = 2r$ renders a highlight with a body reflection component that is less than about 47% of the brightest visible matte point. Accordingly, the highlight line always starts in the upper 53% of the matte line, if $d = 2r$.

Figure 2.9 shows the influence of d on q. Each curve displays q as a function of g for a fixed d. Tables showing the values of $q(g, d)$ can be found in Appendix B. In all curves, q approaches 0.5 as g approaches g_{max}. Note, however, that q is not defined for $g = g_{max}$ where $i_{bLocalMax} = i_{bH} = 90°$, and thus $m_{bLocalMax} = m_{bH} = 0$. The distance d is varied between the curves. For small values of d, the curves exhibit a minimum in the middle of the curve. As d increases, this minimum becomes less pronounced, and q becomes a monotonically decreasing function that approaches 0.5. In the extreme, when $d = \infty$ (orthographic projection), the curve for q assumes the shape of a cosine function for phase angles g under which the globally brightest matte point is visible ($g \leq 90°$), and the shape of a cosecant function, if the brightest matte point is invisible:

$$q_\infty(g) = \begin{cases} \cos(g/2) & \text{if } g \leq 90° \\ 1/(2\sin(g/2)) & \text{if } g \geq 90°. \end{cases} \tag{2.8}$$

A detailed derivation of the formula is given in Appendix C.

The figure and tables indicate that if d is a moderate or large multiple of r ($d \geq 5r$), q does not drop below 0.5. The body reflection component under the highlight then is always at least half as large as the body reflection component at the brightest visible matte point. Under such illumination geometries, the highlight line always starts in the upper 50% of the matte line. Since laboratory set-ups for scenes with several objects in the field of view generally use camera and illumination distances that are larger than five times the object size, we use this property as the 50%-*heuristic* in the segmentation algorithm in Chapter 4.

2.3.3 Spectral Differences between Object Points

In the previous section we have discussed the global properties of spectral variation on an object, as exhibited by the shape of the histogram from an entire object. This section now describes the characteristics of local spectral variation on an object, computed as spectral differences between object points. This analysis provides an understanding of what spectral differences

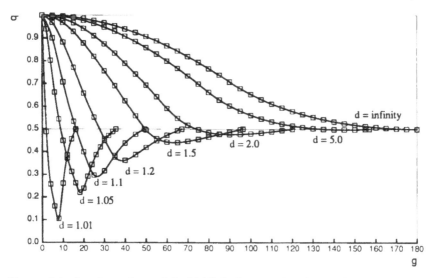

Figure 2.9. Starting point q of the highlight line as a function of phase angle g and distance d.

to expect within a single object and how to distinguish them from spectral changes between objects. It sheds new light on the crosspoint criterion that Rubin and Richards suggested as a means to distinguish spectral changes at material boundaries from spectral changes due to optical effects [157]. It shows that, in the framework of the Dichromatic Reflection Model in which shading and highlight changes can occur simultaneously, the crosspoint criterion is neither a necessary nor a sufficient criterion to determine material changes in images.

According to the Dichromatic Reflection Model, the importance of a spectral vector depends more on its orientation and position in the spectral histogram than on its magnitude. This section discusses the properties of normalized spectral difference vectors. The difference vectors are represented on a sphere, called the *Gaussian spectrum sphere*, in the infinite-dimensional vector space of spectral power distributions, $c(\lambda)$. For convenience, the Figures 2.11-2.13 illustrate the observations on a three-dimensional Gaussian sphere. Note that the notation $\bar{c}(\lambda)$ is used to refer to vectors of unit length, while the lengths of vectors $c(\lambda)$ depend on the magnitude of the spectral power distribution of the reflected light.

In the following, L_{M1} and L_{M2} represent the light reflected from matte pixels P_{M1} and P_{M2} on an object while L_{H1} and L_{H2} refer to light reflected from highlight points, as shown for the cylindrical object in Figure 2.10.

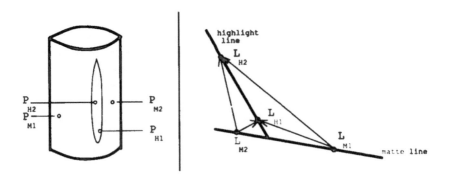

Figure 2.10. Difference vectors between matte and highlight pixels on a skewed T.

The illumination geometry in the figure and the variations in the body and surface reflection components are assumed to be the same as in Figure 2.4. The oval area outlines the highlight pixels. According to equation 2.2 and the discussion on the spectral variation on an object, matte pixels depend only on the body reflection vector, while the spectra of highlight pixels are determined by a nearly constant amount of body reflection, $m_{bH}c_b$[1], and a varying amount of surface reflection:

$$L_{Mi}(\lambda, i, e, g) = m_{bMi}(i, e, g)c_b(\lambda) \tag{2.9}$$

$$L_{Hj}(\lambda, i, e, g) = m_{bH}(i, e, g)c_b(\lambda) + m_{sHj}(i, e, g)c_s(\lambda). \tag{2.10}$$

Equation 2.11 states that the light spectra L_{M1} and L_{M2} reflected from two matte pixels on one object lie on the matte line of the object. Their spectral difference describes the light spectrum of the body reflection component, $c_b(\lambda)$, which indicates the direction of the matte line:

$$L_{M2} - L_{M1} = (m_{bM2} - m_{bM1})c_b(\lambda). \tag{2.11}$$

Since this is the case for any pair of matte pixels, all difference vectors between matte pixels are collinear. On the Gaussian spectrum sphere, all such difference vectors are represented by the same point, $\bar{c}_b(\lambda)$.

Similarly, the orientation of the surface reflection vector is determined by the spectral difference between highlight pixels: Since highlight pixels

---

[1]No index j is given to variable m_{bH} because the amount of body reflection is approximately constant under the highlight.

have an approximately constant matte component, $m_{bH}c_b(\lambda)$, it cancels out when one highlight pixel is subtracted from another. The difference vector describes the light spectrum of the surface reflection component, $c_s(\lambda)$, which determines the direction of the highlight line:

$$L_{H2} - L_{H1} = (m_{sH2} - m_{sH1})c_s(\lambda). \qquad (2.12)$$

All difference vectors between highlight pixels are collinear. They are all represented on the Gaussian spectrum sphere by a single point, $\bar{c}_s(\lambda)$.

The spectral difference between the light from a matte pixel, L_{Mi}, and the light from a highlight pixel, L_{Hj}, defines a difference vector which is a linear combination of $c_b(\lambda)$ and $c_s(\lambda)$. It lies in the dichromatic plane of the object. Its orientation within the dichromatic plane depends on the amounts of body and surface reflection at the two points. Since the differences $(L_{Hj} - L_{Mi})$ and $(L_{Mi} - L_{Hj})$ describe the same difference vector, except for reversing its direction, the subsequent analysis considers only the case of subtracting the matte pixel from the highlight pixel:

$$L_{Hj} - L_{Mi} = m_{sHj}c_s(\lambda) + (m_{bH} - m_{bMi})c_b(\lambda). \qquad (2.13)$$

The equation shows that the difference vector is a linear combination of $c_s(\lambda)$ and $c_b(\lambda)$. Whether the body reflection component is added to the surface reflection component or subtracted from it depends on the sign of $(m_{bH} - m_{bMi})$. If $m_{bH} > m_{bMi}$, the difference vector results from adding some amount of $c_b(\lambda)$ to $c_s(\lambda)$. If, on the other hand, $m_{bH} < m_{bMi}$, the difference vector is determined by subtracting some amount of $c_b(\lambda)$ from $c_s(\lambda)$. Accordingly, the unit circle representing the difference vectors in the dichromatic plane can be divided into zones of vector addition and vector subtraction, as shown in Figure 2.11. The zones formed by vector addition are called *inside zones*, since they lie between $\bar{c}_s(\lambda)$ and $\bar{c}_b(\lambda)$ or between $-\bar{c}_s(\lambda)$ and $-\bar{c}_b(\lambda)$. Zones formed by vector subtraction are called *outside zones*.

The dichromatic plane intersects the Gaussian spectrum sphere in a great circle which passes through the points $\bar{c}_b(\lambda)$ and $\bar{c}_s(\lambda)$. Hence, all difference vectors between matte and highlight pixels, when normalized to unit length, form a great circle on the Gaussian spectrum sphere. Normalized difference vectors that were generated by vector addition are represented by points on the inside segment of the great circle, while normalized difference vectors resulting from vector subtraction lie on the outside segment beyond $\bar{c}_s(\lambda)$. This is shown in Figure 2.12. The difference vectors $(L_{H1} - L_{M1})$ and $(L_{H2} - L_{M1})$ from Figure 2.10 are formed by vector addition. They are

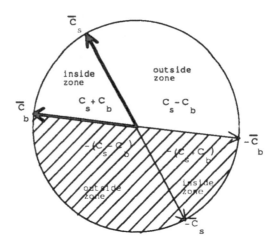

Figure 2.11. Difference vectors on the dichromatic plane.

represented on the Gaussian spectrum sphere by points that lie between $\bar{c}_b(\lambda)$ and $\bar{c}_s(\lambda)$. The difference vectors $(L_{H1} - L_{M2})$ and $(L_{H2} - L_{M2})$, on the other hand, result from subtracting some amount of $c_b(\lambda)$ from $c_s(\lambda)$. These vectors point to positions beyond $\bar{c}_s(\lambda)$ on the great circle.

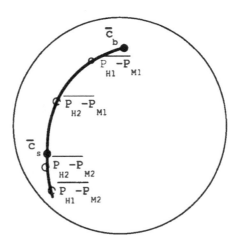

Figure 2.12. Orientations of spectral differences on one object, shown on the Gaussian spectrum sphere.

When there are two or more objects with highlights in the scene, each of them generates a partial great circle on the Gaussian spectrum sphere. Since all highlight clusters are parallel to one another, the segments all meet or intersect at the point representing the surface reflection vector, $\bar{c}_s(\lambda)$, as shown in Figure 2.13. In addition to analyzing spectral differences within each object, spectral differences across material boundaries, i.e., between points on two neighboring objects O_1 and O_2, must also be modeled. Depending on whether the difference is computed between two matte pixels $L_{M(O1)}$ and $L_{M(O2)}$, two highlight pixels $L_{H(O1)}$ and $L_{H(O2)}$, or between a matte and a highlight pixel, $L_{M(Oi)}$ and $L_{H(Oj)}$, the difference vector is determined by equation 2.14, 2.15, or 2.16:

$$
\begin{aligned}
L_{M(O2)} - L_{M(O1)} &= m_{b(O2)}c_{b(O2)}(\lambda) - m_{b(O1)}c_{b(O1)}(\lambda) \quad &(2.14) \\
L_{H(O2)} - L_{M(O1)} &= m_{bH(O2)}c_{b(O2)}(\lambda) - m_{b(O1)}c_{b(O1)}(\lambda) \\
&\quad + m_{s(O2)}c_s(\lambda) \quad &(2.15) \\
L_{H(O2)} - L_{H(O1)} &= m_{bH(O2)}c_{b(O2)}(\lambda) - m_{bH(O1)}c_{b(O1)}(\lambda) \\
&\quad + (m_{s(O2)} - m_{s(O1)})c_s(\lambda). \quad &(2.16)
\end{aligned}
$$

Equations 2.14 through 2.16 describe spectral variation across material boundaries as a function of the spectra of the two body reflection components of neighboring objects and – if highlight pixels are involved in the differencing operation – of the (unique) highlight reflection component. The orientations of the difference vectors are thus confined to a two- or three-dimensional subspace of the infinite-dimensional vector space. Equation 2.14 shows that the difference between the body reflection components is formed by vector subtraction. Accordingly, if no highlight component exists, the difference vector lies on the outside segment next to $\bar{c}_{b(O2)}(\lambda)$ on the great circle passing through $\bar{c}_{b(O1)}(\lambda)$ and $\bar{c}_{b(O2)}(\lambda)$. This is shown in Figure 2.13 as the dark line segment going downwards from $\bar{c}_{b(O2)}(\lambda)$. If a highlight component does exist, it is added to the matte difference. The difference vector then lies in the three-dimensional subspace of the Gaussian spectrum sphere that extends from the above described outside segment of the great circle through $\bar{c}_{b(O1)}(\lambda)$ and $\bar{c}_{b(O2)}(\lambda)$ toward $\bar{c}_s(\lambda)$ and that is bounded by the great circles of the dichromatic planes of the two objects. The corresponding area in Figure 2.13 is shaded.

This analysis sheds a new light on previous investigations on spectral changes in images [157]. It shows that the crosspoint criterion developed by Rubin and Richards is not useful in the framework of the Dichromatic Reflection Model: In search of a criterion to distinguish spectral changes at material boundaries from spectral changes due to physical events such as shading, shadows and highlights, Rubin and Richards analyzed the basic

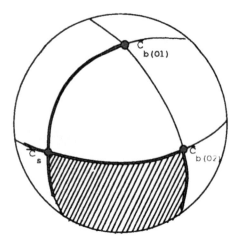

Figure 2.13. Orientations of spectral differences on two objects, shown on the Gaussian spectrum sphere.

influences of such events on the reflected light. They observed that none of the events caused the light spectra of neighboring object points to cross each other, if the physical events occurred separately from each other, i.e., if every spectral change between neighboring object points was related to only a single physical event. They suggested that a spectral crosspoint for neighboring object points was a sufficient criterion to indicate a material change, although not a necessary one, since the light reflected from neighboring objects may have non-intersecting spectra.

In the terminology of spectral difference vectors in the infinite-dimensional vector space, the crosspoint criterion tests for a *negative difference vector*, a vector for which the difference between the spectral curves of the neighboring points is negative at some but not all wavelengths.[2] The Dichromatic Reflection Model relaxes Rubin's and Richard's assumption that any spectral change is caused by a single physical event. Instead, the model allows for changes both in body reflection (shading) and surface reflection (highlights) between neighboring object points. Equation 2.13 formally describes the resulting spectral change between a matte and a highlight point on an object. As discussed above, the difference vector is formed by subtracting some amount of body reflection from the surface re-

---

[2] *Positive difference vectors*, on the other hand, represent spectral curves without crosspoints, such that the spectral difference between the two curves is either positive at all wavelengths or negative (in which case the vector can be reversed into all positive components).

flection component, if $(m_{bH} - m_{bMi}) < 0$. Whether the resulting difference vector then has a positive or a negative direction depends on the spectral curves of $c_s(\lambda)$ and $c_b(\lambda)$. If, for example, $c_s(\lambda_i) = 0$ and $c_b(\lambda_i) > 0$ at some wavelength λ_i, the spectral difference at λ_i is negative, while another wavelength, λ_j, may exist at which $c_s(\lambda_i) > 0$ and $c_s(\lambda_i) = 0$, resulting in a positive spectral difference. The spectral difference between points L_{H1} and L_{M2} in Figure 2.10 is an example of a negative difference vector. This difference vector proves that, under the assumptions of the Dichromatic Reflection Model, negative spectral differences and thus spectral crosspoints can occur at highlights. Consequently, the spectral crosspoint criterion is neither a necessary nor a sufficient criterion to determine a material change between objects on which the body reflection component varies independently of (and generally simultaneously with) the surface reflection component.

2.4   Dimensionality of the Measurement Space

The discussion in the previous section indicates that, within the Dichromatic Reflection Model, a two-dimensional space (the dichromatic plane) suffices to describe spectral differences on a single object. The same is the case for spectral differences between matte pixels on neighboring objects. As shown in Figure 2.13, such spectral variations fill different two-dimensional subspaces in a higher-dimensional space. A higher-dimensional (at least three-dimensional) space is necessary to identify the two-dimensional structure of such spectral variations and to distinguish between different variations on and between different objects.

In general, spectral differences between highlight pixels or a matte and a highlight pixel on two objects also have to be considered. Such differences are confined to a three-dimensional space, and a space of at least four dimensions would be needed to detect their structure. However, such differences are rare occurrences in most images and may be ignored for most purposes.

2.5   Material Classes

Although non-uniform, opaque dielectrics represent a large class of materials, some others exist as well. This section provides a classification and a short overview of materials and their characteristic appearances. It points out why the Dichromatic Reflection Model is not applicable to other material classes. A summary of the classification is shown in Figure 2.14.

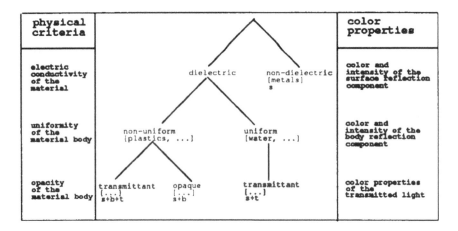

Figure 2.14. Classification scheme for materials.

An important criterion for classifying the appearance of materials is the
electric conductivity of the material. This distinction separates dielectric
materials from non-dielectric (conducting) materials, such as metals Metallic objects have a very high surface reflectance. At normal incidence, for
example, silver and aluminum reflect over 90% of all visible light, while
dielectrics generally reflect only about 4% directly from the material surface [80]. The high reflectance power of metals is related to their ability to
conduct electricity and expressed in the extinction coefficient. It is coupled
with high absorption. The surface reflection component of metals is generally very strong, and its spectral distribution depends on the material, as
can be seen in the yellow highlight on golden objects. The body reflection
component of metals, on the other hand, is negligible since practically all
light that enters the body is absorbed. The result is that only surface reflection can be seen on metallic objects. Accordingly, light reflection from
metals depends mainly on a single reflection process (surface reflection)
and can be roughly described by a single spectral vector, $c_s(\lambda)$. This is
indicated by the symbol s in Figure 2.14. If a metallic object is illuminated by a single light source, its spectral cluster thus forms a line in the
spectral space. In contrast, the Dichromatic Reflection Model describes
light reflection from (non-uniform, opaque) dielectrics as a function of two
independent reflection processes (body and surface reflection), as indicated
by the $s + b$ in Figure 2.14.

For dielectrics, a further classification criterion separates materials with
uniform material bodies from materials with non-uniform bodies. This

criterion is of little relevance for non-dielectric materials since the body reflection of such materials is negligible. The decision as to whether materials
are uniform or non-uniform is a question of scale. How large can the basic
element be? Different research communities have established classifications
using different scale thresholds. In relation to the discussion of scattering
and absorption in the material body, this book relates uniformity to the
wavelength of visible light: Dielectrics are considered to be non-uniform,
if the size of the particles is large compared to the wavelengths of visible
light. A similar distinction is common in physics, where the analysis of
light reflection from rough surfaces is described by geometric optics when
the surface roughness is coarse compared to wavelength and by physical
optics (wave-theory) otherwise. Examples of uniform dielectrics are glass,
clear water, perfect gemstones and some crystals such as quartz. If a dye is
dissolved in a solution, it is also a uniform dielectric material according to
the above classification. Examples of non-uniform dielectrics are materials
with pigments larger than the wavelength of visible light, such as paints,
plastics, ceramics and paper. This class also includes dyed textiles, since
the size of the fibers is large compared to the wavelengths under consider-
ation. Other literature [163] has used the term "inhomogenous material"
in similar contexts. Material homogenuity is a misleading criterion [86]
because a material with particles smaller than the wavelengths of visible
light still has to be considered to be inhomogeneous, although is behaves
uniformly.

Since uniform materials do not contain any particles large enough to
scatter light, no light beam is reflected back to the material surface from
where the light entered. As a result, no body reflection exists for such
materials: Any light entering the material body is either absorbed or exits
at the opposite side. This leads to considering a third process by which
light interacts with matter: *light transmission*. It determines the opacity,
translucency, or transparency of materials. A material is *opaque* if no part
of the refracted light beam is able to penetrate through the entire material.
Otherwise, the material is *transmittant*, appearing either *transparent* or
translucent, depending on how much the light is scattered in the material
body and at the material surfaces due to a roughened surface. The opacity
of a material depends on the thickness of the material, on its absorption
power, and on the distribution of the scattering particles – especially on
the distance between them.

Body reflection and transmittance properties divide dielectrics into three
classes: uniform dielectrics, non-uniform transmittant dielectrics, and non-
uniform opaque dielectrics. The first class, uniform dielectrics, do not exhibit body reflection but only transmit light since the light is not scattered
and reflected back in the material body. Spectral variation on uniform

dielectrics is thus determined by light transmission and by surface reflection, as indicated by $s + t$ in Figure 2.14. The second class, non-uniform dielectrics, exhibit spectral variation that depends on surface reflection, body reflection, and transmittance, as indicated by $s + b + t$ in Figure 2.14. For the third class, opaque, non-uniform dielectrics, light transmission does not exist, and spectral variation on the object is governed only by body and surface reflection. This is indicated by $s + b$ in Figure 2.14. In conclusion, spectral variation on uniform dielectrics, as well as on non-uniform, opaque dielectrics is governed by two processes, while spectral variation on non-uniform, transmittant dielectrics depends on three processes. Transmitted light can only be seen by a camera, however, if the light source is behind the object, while body and surface reflection need a light source in front of the object. Thus, since the Dichromatic Reflection Model assumes that only one light source exists in the scene, light transmission cannot be observed in this model in combination with body and/or surface reflection at an object point. Accordingly, if the light source is behind an object area, spectral variation depends only on a single spectral vector which is determined by the transmitting properties of the material and the spectral power distribution of the illumination. The result is a linear spectral cluster. If the light source is in front of the object, on the other hand, opaque and transmitting non-uniform materials exhibit spectral variation that is a linear combination of the body and surface reflection vector. Uniform dielectrics only exhibit surface reflection, resulting in a linear spectral cluster.

Summarizing the discussion on spectral variation seen on the various material classes under illumination from a single light source, two-dimensional spectral variation only occurs on non-uniform dielectrics that are illuminated from the front. Such spectral variation is determined by the surface and body reflection process. The Dichromatic Reflection Model is suitable to describe spectral variation on such materials. All other material types exhibit one-dimensional spectral variation which may depend on any of the three processes, under varying physical properties. Accordingly, two-dimensional spectral variation may be taken as a strong clue for identifying non-uniform dielectrics and for relating the variation to shading and highlights. Linear spectral variation, on the other hand, can be the result of a number of physical scene properties and thus cannot be interpreted unambiguously.

The above classification scheme provides an appropriate distinction between many common materials, although it does not cover all existing materials. Light reflection from some materials depends on phenomena that have not been a part of the above criteria. For example, the classification scheme does not discuss fluorescent materials. Furthermore, some dielectric materials are known to exist that look like metals. Their surface reflection component has a metal-like appearance – an effect known as bronzing. It is related to interference occurring if a dielectric is covered by a very thin layer or film[85]. Further complex effects can appear on pigmented dielectrics, if they contain metallic flakes, if the pigments have a preferred orientation in the material body, or if they are protruding from the medium. These and other effects are described in [82].

2.6   Summary

This chapter has presented a physical reflection model. The model describes light reflected from dielectric, non-uniform, opaque materials as a mixture arising from shading (body reflection) and highlights (surface reflection) on the object. When combined with a geometric analysis of shading and highlight variation on objects, the model indicates that light reflection on a single, uniformly colored object varies in a two-dimensional subspace – a dichromatic plane – of the infinite-dimensional space of vector space of spectral power distributions. Within the plane, the light mixtures from different object points form a spectral cluster that looks like a skewed T. According to quantitative investigations on how light reflection varies on a sphere, the linear branches of the skewed T generally meet in the upper 50% of the horizontal bar of the T (the matte line).

The model is applicable to a certain class of materials: non-uniform, opaque dielectrics, such as plastics, paints, papers, ceramics, and textiles. It is not applicable to non-dielectrics, such as metals. Neither does it describe light variation from uniform and/or transmitting dielectrics, such as water or dyes. The description of the spectral clusters also assumes that the objects in the scene are uniformly colored and that there is only a single light source.

3

A Sensor Model

Chapter 2 described light reflection in a theoretical, physical model, but the measured data in a recorded image is influenced by the characteristics of the camera. This section covers the influence of such camera characteristics on the image data. Since some of these influences disturb the light reflection properties stated in the Dichromatic Reflection Model, it is necessary to provide methods that restore the physical properties of light reflection. Where this is impossible, image analysis algorithms must be able to detect or tolerate the inaccuracies in the image data. The images shown in this book were taken in the Calibrated Imaging Laboratory [162] at Carnegie Mellon University using a black-and-white CCD-camera[1] and spectral filters[2]. This chapter describes the problems occuring with such cameras and demonstrates the influence of shading, highlights, and camera limitations on pixel values in real images.

---

[1]SONY AVC-D1.

[2]Wratten filters with numbers 25, 58 and 47 and a Corion FR-400 filter to suppress infrared light.

3.1   Spectral Integration

The Dichromatic Reflection Model describes light reflection using a physically precise representation: the continuous light spectrum. However, the
human eye as well as light sensing devices use a finite set of samples to
describe the spectrum. Sample measurements are obtained by filtering the
light spectrum and integrating over the filtered spectrum. This process
is called *spectral integration*. It integrates the amount of incoming light,
$L(\lambda)$, weighted by the spectral transmittance, $\tau_f(\lambda)$, of the respective filter
f and the spectral responsivity of the camera, $s(\lambda)$, over all wavelengths λ
to obtain pixel value C_f:

$$C_f = \int L(\lambda, i, e, g)\tau_f(\lambda)s(\lambda)d\lambda. \tag{3.1}$$

The color images shown in this text have been obtained using red, green
and blue spectral filters, reducing the infinite-dimensional vector space to
a three-dimensional space, called the *color space*. The spectrum of an
incoming light beam at position (x, y) is represented by a triple, called a
color pixel, $\mathbf{C}(x, y) = [C_R, C_G, C_B]$, where x and y are determined by the
photometric angles i, e and g used in the Dichromatic Reflection Model
and by the position of the object relative to the camera. For convenience,
$[C_R, C_G, C_B]$ will also be referred to as $[R, G, B]$.

Spectral integration from the infinite-dimensional vector space of spectral
power distributions to a three-dimensional color space is a linear transformation [48, 160]. Thus, if a light beam $L(\lambda)$ is a mixture of two lights,
$L_1(\lambda)$ and $L_2(\lambda)$, in the infinite color space, then the color vector $\mathbf{C}$, resulting from spectral integration on $L(\lambda)$ is the same linear combination of
the color vectors $\mathbf{C_1}$ and $\mathbf{C_2}$ that result from spectral integration of $L_1(\lambda)$
and $L_2(\lambda)$. For this reason, the linear relationship between reflected light
and the colors of surface and body reflection, as stated in equation 2.2,
is maintained under spectral integration. This leads to the Dichromatic
Reflection Model for the three-dimensional color space. It is expressed by
equation 3.2:

$$\mathbf{C}(x, y) = m_s(i, e, g)\mathbf{C_s} + m_b(i, e, g)\mathbf{C_b}. \tag{3.2}$$

Equation 3.2 describes the color pixel value $\mathbf{C}(x, y)$ as a linear combination of the two color vectors, $\mathbf{C_s} = [R_s, G_s, B_s]$ and $\mathbf{C_b} = [R_b, G_b, B_b]$,
which describe the colors of surface and body reflection on an object in the
scene. Within the three-dimensional color space, $\mathbf{C_s}$ and $\mathbf{C_b}$ span a dichro-

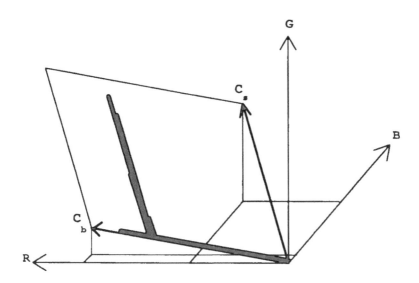

Figure 3.1. Color cluster in a three-dimensional color space.

matic plane which contains a parallelogram in which the color clusters lie. This is shown in Figure 3.1.

Note that, in principle, spectral integration introduces a nonuniqueness, called *metamerism* [83, 87], in the resulting finite representation: The infinite-dimensional color space is partitioned into a set of equivalence classes, each of which contains an infinite number of spectral curves that are all represented by the same color. This is irrelevant to the Dichromatic Reflection Model, however, since the finite color space maintains the linear relationships between the reflected light mixtures from one object.

3.2 Limited Dynamic Range

Real cameras have only a limited dynamic range to sense the brightness of the incoming light. This restricts the analysis of light reflection in real color images to a *color cube*, as shown in Figure 3.2. Its walls denote the upper and lower limits of the dynamic range in the respective color bands.

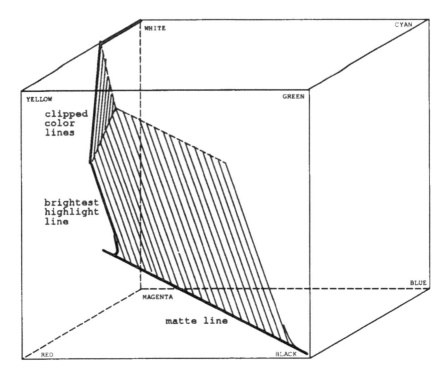

Figure 3.2. Color cluster in the color cube, with color clipping.

3.2.1   Color Clipping

If the incoming light is too bright at some pixel position and exceeds the dynamic range of the camera, the camera cannot sense and represent it adequately. Such *color clipping* may occur in one, two, or all three color bands, depending on the color and intensity of the light. It can be a problem for measuring the color of highlights or bright objects. In the color histograms, clipping causes the clusters to bend along the walls and edges of the color cube (see Figure 3.2). In a very bright area, the dynamic range of the camera may be exceeded in all three color bands, resulting in white pixels even though the color of the light incident at the camera may not be white.

Such *clipped color pixels* do not follow the characteristics of the Dichromatic Reflection Model. They form clipped color clusters that generally point in a different direction than the body and surface reflection vectors of an object. They generally do not even lie in the dichromatic plane. The direction of a clipped color cluster is not arbitrary, however, but is well-defined with respect to the orientations of the body or surface reflection vectors. Let $\mathbf{C_b} = (r_b, g_b, b_b)$ and $\mathbf{C_s} = (r_s, g_s, b_s)$ be the body and surface reflection vectors of an object, $\overline{\mathbf{C}}_b = (\overline{r}_b, \overline{g}_b, \overline{b}_b)$ and $\overline{\mathbf{C}}_s = (\overline{r}_s, \overline{g}_s, \overline{b}_s)$ the corresponding unit vectors, and $\overline{n} = (\overline{r}_n, \overline{g}_n, \overline{b}_n) = \overline{\mathbf{C}}_b \times \overline{\mathbf{C}}_s$ the normal to the dichromatic plane, with $\overline{r}_n = (\overline{g}_b\overline{b}_s - \overline{b}_b\overline{g}_s)/norm$, $\overline{g}_n = (\overline{b}_b\overline{r}_s - \overline{r}_b\overline{b}_s)/norm$, $\overline{b}_n = (\overline{r}_b\overline{g}_s - \overline{g}_b\overline{r}_s)/norm$, and $norm$ is the sine of the angle between $\overline{\mathbf{C}}_b$ and $\overline{\mathbf{C}}_s$.

If the clipped cluster is a clipped extension of the highlight cluster, the clipped vector $\overline{\mathbf{C}}_c = (\overline{r}_c, \overline{g}_c, \overline{b}_c)$ has the same (although renormalized) components as the surface reflection vector[3], except for the clipped component, which is zero. To demonstrate the properties of color clipping, the following discussion assumes that the red component is clipped, $\overline{r}_c = 0$. The orientation of the clipped vector then is

$$\overline{\mathbf{C}}_c = \frac{(0, \overline{g}_s, \overline{b}_s)}{norm}, \tag{3.3}$$

with

$$norm = \sqrt{\overline{g}_s^2 + \overline{b}_s^2}. \tag{3.4}$$

If the clipped vector was to lie in the dichromatic plane, its dot product with the dichromatic normal would have to be zero:

$$\overline{\mathbf{C}}_c \cdot \overline{n} = 0. \tag{3.5}$$

Componentwise expansion of the dot product (with $\overline{r}_c = 0$) yields

$$\overline{g}_s\overline{g}_n + \overline{b}_s\overline{b}_n = 0. \tag{3.6}$$

Expressing $\overline{n}$ as the cross product of $\overline{\mathbf{C}}_b$ and $\overline{\mathbf{C}}_s$ leads to

$$\overline{g}_s(\overline{b}_b\overline{r}_s - \overline{r}_b\overline{b}_s) + \overline{b}_s(\overline{r}_b\overline{g}_s - \overline{g}_b\overline{r}_s) = 0, \tag{3.7}$$

which can be reduced to

$$\overline{r}_s(\overline{g}_s\overline{b}_b - \overline{b}_s\overline{g}_b) = 0. \tag{3.8}$$

Since $\overline{g}_s\overline{b}_b - \overline{b}_s\overline{g}_b$ describes the red component of $\overline{n}$, $\overline{r}_n$, equation 3.8 simplifies to

$$\overline{r}_s\overline{r}_n = 0. \tag{3.9}$$

---

[3] A similar formalism holds if the clipped cluster is an extension of the matte cluster.

Equation 3.9 states that the clipped cluster will only lie in the dichromatic plane if the constraint in the clipped component (here, $\bar{r}_c = 0$) is naturally provided by the product $\bar{r}_s \bar{r}_n$. This means that either the dichromatic plane intersects the clipping wall at a right angle (i.e., $\bar{r}_n = 0$, and the dichromatic normal is parallel to the wall) or that the highlight cluster runs parallel to the wall ($\bar{r}_s = 0$), in which case no clipping takes place. In general, the above condition does not hold. Therefore, the clipped cluster generally does not lie in the dichromatic plane.

This analysis shows that there exists a strong relationship between the orientation of the highlight (or matte) cluster and the clipped cluster. It indicates that the orientation of the clipped cluster can be used partially to constrain the orientation of the surface (or body) reflection vector. Exploiting this constraint is a subject for future work.

3.2.2 Blooming

A further influence of the limited dynamic range on the measured pixel values is *blooming* in a CCD-camera [10]. With CCD-cameras, too much incident light at a pixel may completely saturate the sensor element at that position [19]. As a result, more charges are generated at a pixel than the sensor element can hold, the excess carriers spread then out to the adjacent pixels and change their values in proportion to the magnitude of the overload. In the case of very bright spots in the scene, blooming may travel quite far from the originally overloaded sensor element and can cause extended white areas in an image. Pixels which have increased color values because of blooming are called *bloomed color pixels*. Their color values may be increased arbitrarily in one, two or all three color bands.

Both color clipping and color blooming are problems for image sensing with most currently available cameras, especially for the sensing of scenes with highlights. They interact badly with another effect that is common in CCD-cameras: Spatial averaging between neighboring pixels that is commonly built into CCD-cameras to reduce the variance in the sensing characteristics in the array. Because of such spatial averaging, color clipping and blooming is "smoothed" over an image area. As a consequence, these effects cannot be detected by merely thresholding the color image at maximal intensity: Because of the averaging, the effect is spread to neighboring pixels, increasing their values and decreasing the values of the clipped pixels. In the images in this book, color values in the upper 10% of the intensity scale are suspected to be influenced by color clipping or blooming.

3.3   Color Balancing and Spectral Linearization

This section describes two methods that are aimed at providing color data that maintains the linear relationships (skewed T shape of color clusters) described in the Dichromatic Reflection Model. When a CCD-camera takes a picture, the measured color pixel values depend on the responsivity of CCD-cameras to varying wavelengths and incident flux. Unfortunately, such cameras do not maintain a constant sensitivity over the visible spectrum and over a wide range of incident flux. A typical curve of the spectral responsivity of CCD-cameras is shown in Figure 3.3 [19]. Because of the ramp-like spectral responsivity of CCD-cameras within the visible spectrum, the blue band registers much less contrast than the red band. This lowers the signal to noise ratio in the blue band. In order to provide an equal scaling on the three color axes in the color space, the pixel data needs to be rescaled separately in the color bands. This procedure is called *color balancing*. A simple multiplication of the color data by constant factors, however, does not improve the dynamic range in the color bands and thus does not improve the quality of the color data; it only introduces artificial gaps into the color distribution in the color histogram. Instead, the images shown in this book were color balanced by controlling the camera aperture during the picture-taking process. The apertures for green and blue exposures under tungsten light were $\frac{1}{2}$ and $1\frac{1}{2}$ f-stops (respectively) lower than the aperture used for the red exposure. The upper left quarter of Color Figure 4 (see color insert) shows the image of a Macbeth Color Checker color chart that was obtained using color balancing by aperture control. The lower left quarter shows an image of the same color chart taken without color balancing. The unbalanced image looks more orange, since the blue and green color bands do not contribute appropriately to the pixel colors. For this reason, the grey scale in the lowest row of the chart is brown. The grey scale of the color balanced image, on the other hand, is approximately grey. The pictures shown hereafter have been color balanced by aperture control.

Note that CCD-cameras are very sensitive to infrared light. Since most color filters in the visible spectrum are nearly transparent in the infrared, camera responsivity to the infrared range can add a significant amount of measured intensity to the color values. As a result, the colors in the image can be "washed out". To eliminate the influence of infrared light, all images in this book have been taken with a total IR suppressor[4] in front of the camera.

---

[4] Corion FR-400.

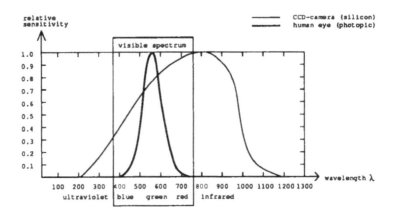

Figure 3.3. Spectral responsivity of CCD-cameras and the human eye.

The color pixels also depend on camera response to incident light flux. Most modern sensors have a linear response to flux [19]. However, since the luminous output of display devices is related by a power law to the driving voltage, $L = kV^\gamma$, it is common practice in the television industry to produce a signal proportional to $Y^{1/\gamma}$ in the camera, with 2.2 as a typical value for γ [171]. This is known as *gamma-correction*. As a result, the camera output is related by an inverse power law to the incident flux. The diagrams in Color Figure 4 show this relationship for the camera that was used for taking the color images in this book. To measure this relationship, the known reflectance values of the grey scale of the Macbeth color chart were compared with the measured, averaged intensity values in the corresponding image areas in the three color bands. The upper diagram in Color Figure 4 displays the measurements from a color image that was taken with color balancing by aperture control. The horizontal axis represents the known reflectance values of the grey blocks. The vertical axis represents the measured intensities in the three color bands. The points in the diagram approximately lie on the typical curve for gamma-correction. The lower diagram displays the same measurements from the unbalanced color image. It demonstrates the decreasing spectral responsivity of cameras to short wavelengths.

Color Figure 5 (see color insert) demonstrates how gamma-correction disturbs the linear properties of the Dichromatic Reflection Model. The upper left quarter shows the image of an orange cup. The cup does not exhibit a significant amount of surface reflection in the camera direction. Consequently, all pixels in the outlined image area are matte pixels and should form a straight matte line in the color space. The upper right quarter of Color Figure 5 demonstrates that this is not the case: Gamma-correction introduces curvature into the color space, distorting the linear properties of the Dichromatic Reflection Model.

The color images in this book are spectrally linearized by fitting separate interpolating cubic splines to the measured responsivity data in each color band, as suggested by LeClerc [106]. To model very bright light reflection that can occur in highlights, the linearization function is rescaled and extrapolated from the brightest measurement outward by a square root curve. The rescaling is necessary because linearizing very bright spots in the original image produces values beyond the representable range of pixel values, due to the slope of the fitted curve. These functions are then used to generate a look-up table which relates the measured intensities in every color band to the incident flux. The upper right box of Color Figure 6 (see color insert) shows the red, green and blue cubic splines that were fit to the responsivity data of Color Figure 4. The lower left box of Color Figure 6 displays the linearized image of the color chart. When compared with the unlinearized image (upper left box), the colors of the linearized image look more saturated. The image is also darker, due to the rescaling of the intensities. The lower half of Color Figure 5 shows that when the linearization method is applied to the image of the orange cup, the resulting color histogram is approximately linear, pointing in the direction of the body reflection vector. All pictures shown hereafter are linearized.

3.4   Chromatic Aberration

A further influence on the measured pixel values is caused by *chromatic aberration* in the camera lens. The refractive index of optical glass is a function of wavelength, with shorter wavelengths (e.g., blue light) being refracted more strongly than longer wavelengths (e.g., red light). This principle is known as color dispersion and has been characterized by Fraunhofer as a set of monotonically decreasing curves that describe the refractive index as a function of wavelength [80].

Since the focal length F of the camera lens depends on the refractive index of the optical glass in the lens, F is also a function of wavelength λ, with F increasing with λ. As a consequence, the optimal position of the image plane in the camera also changes with wavelength, as shown in Figure 3.4. Consequently, it is impossible to generate perfectly precise color images from a single camera setting. If the focus is perfectly set for the red

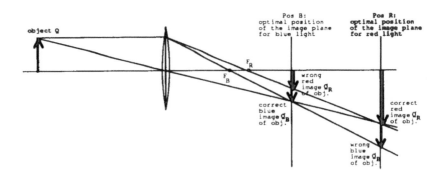

Figure 3.4. Chromatic aberration.

image band, the green and the blue bands (even more so) show magnified projections of the objects. If, on the other hand, the focus is correct for the blue image band, the objects are shrunk in the green and red bands. Furthermore, since those color bands are out of focus, the object edges are blurred. For optimal results, the focal length of the camera needs to be set separately for the red, green and blue band. This, however, is very hard to achieve manually. A future extension to the Calibrated Imaging Laboratory is planned to provide the necessary focus and magnification control in an automatic camera control system [145].

Figure 3.5 shows that the difference in the positions of object points in the color bands is small at the image center and increases toward the image boundaries: The object point M on the optical axis is projected along the optical axis into the center of the image. Since F_R and F_B both lie on this projection ray, there is no color dispersion. As the distance from the optical axis increases, however, the difference in focal lengths has more effect on the projection geometry.

Color Figure 7 (see color insert) shows how chromatic aberration influences the colors in images. The upper left quarter of the figure shows the color picture of a black-and-white calibration grid which has been taken by focussing on the red color band. The color pixel values are sampled along short horizontal intervals within a scanline in the middle of the image. The first interval is taken across the leftmost black vertical bar. The second interval is taken across the black vertical bar at the image center, and the third interval runs across the rightmost bar. The intervals are shown in white on the original image. The upper right quarter and both lower quarters of Color Figure 7 display the intensity profiles in the red, green and blue color bands along the intervals. All profiles have high intensity values

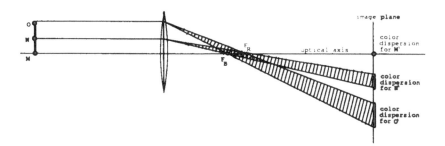

Figure 3.5. Chromatic aberration as a function of the position in the scene.

at the beginnings and ends of the curves, corresponding to white pixels, and low intensity values in the center, where a black bar occurs. The lower left quarter shows the red, green and blue profiles of the center interval. The curves of all three color bands are very similar, dropping to dark intensities at the same pixels. The upper right quarter shows the profiles from the left interval. In this image area, the positions of the intensity drops do not coincide in the three color bands, the blue profile dropping to the left of the green and red profile. The lower right quarter presents the opposite behavior in the right interval. This shows that the grid is slightly magnified in the blue color band when compared to the red band.

Chromatic aberration has a strong influence on the color pixel values along edges in the outer image areas. Since the color bands do not increase or decrease simultaneously, pixels in the rising and falling parts of the profiles generally exhibit color changes that have little relationship to the physical processes in the scene. The color changes do, however, exhibit systematic color aberration along the edge and form distinctive features in the color space. This is a serious problem for the measurement and analysis of highlights in the outer image areas. How much chromatic aberration distorts the color values depends on the quality of the camera lens. The distortion appears quite strongly in some of the images in this book.

3.5   Examples: Color Clusters from Real Color Images

This section demonstrates how shading, highlights and camera limitations influence the pixel values in real color images. It shows and discusses a series of color images that were taken in the Calibrated Imaging Laboratory of a variety of dielectric materials (plastics, paper and ceramics) under several illumination colors. Black curtains on the walls and a black ceiling were used to eliminate indirect illumination in the visible spectrum of the scene.

Color Figure 8 (see color insert) shows the image of a green, a yellow and an orange plastic cup, illuminated by yellow light. Color Figure 9 (see color insert) shows the color histogram that results from projecting the color pixels from the outlined areas of Color Figure 8 into the color cube. The color histogram consists of several linear clusters. There are three linear clusters which start close to the black corner of the color cube. These clusters contain the matte pixels of the three cups. They lie along the orange, yellow and green matte lines of the cups. The matte clusters are connected to linear highlight clusters which lie along the respective highlight lines of the cups. For the yellow cup, the matte and highlight clusters are nearly collinear because the material and illumination color are nearly the same. Note that the highlight clusters from all three cups are approximately parallel to each other. Furthermore, they are parallel to the color vector describing the yellow illumination because the refractive indices of the cups are fairly constant over the visible spectrum. The figure thus supports the assumption of the Dichromatic Reflection Model that surface reflection from dielectric materials has the same color as the illumination. The figure demonstrates that this property of dielectrics is readily detectable in real images and that the direction of the highlight clusters can be used to determine the illumination color. This is useful for color constancy algorithms [33, 109, 119], which try to remove the influence of the illumination color from the body reflection component, "normalizing" the image to a standard white illumination.

Further linear clusters in the color histogram contain clipped color pixels. These are displayed in Color Figure 9 in their complementary color (blue). Such color clipping is identified through a high threshold on the measured intensities in the three color bands. The clipped color pixels come from the centers of the highlights where the reflected light exceeds the dynamic range of the camera. Accordingly, the centers of the highlights contain white pixels, whereas the pixels closer to the highlight boundaries exhibit yellow surface reflection.

Color Figures 10 and 11 (see color insert) demonstrate how blooming influences the color distribution in the color histogram. The figures show an image and color histogram of an orange cup under yellow light. The highlight center on the cup is so bright that charges spread out to adjacent pixels, increasing their values. Because the cup is orange and the illumination is yellow, blooming occurs mainly in the red color band. The bloomed color pixels thus have an increased red color value, while the green and blue values are correct. As a result, the highlight cluster is spread out along the red axis of the color cube. The same effect is present – to a smaller degree

– in the highlight cluster of the green cup in Color Figure 9. Since blooming cannot always be completely avoided in pictures with highlights, it has
to be accounted for in the color histograms. Under the assumption that
blooming occurs only at a few pixels in the image and that it increases the
color pixel values arbitrarily, it is unlikely that such color pixel values occur
often in the image. Accordingly, colors with a low frequency count in the
color histogram may be suspected to be caused by blooming or other noise
(*blooming heuristic*). Color Figure 11 displays such color pixels in white.

   Color Figures 12 and 13 (see color insert) display the influence of chromatic aberration on color images and color clusters for the example of the
green plastic donut in the lower right corner of the image of the plastic
scene (Color Figure 1). Color Figures 12 and 13 exhibit two different views
of the color histogram of the green donut. Color Figure 12 displays the
histogram in a side view of the cluster, while Color Figure 13 shows it in
a top view. They show that the highlight branch of the color histogram
spreads very widely. The spread of the highlight branch occurs within a
plane which approximately passes through the yellow and cyan corners of
the color cube. Within the plane, most pixels lie around the center line,
representing the "true" white color direction of the highlight cluster. However, a significant number of pixels show a rather large deviation from the
center line toward yellow/orange or cyan/blue colors. In relation to the
green body vector and the white surface reflection vector of the donut, the
orange pixels contain too much red and not enough blue color. Such a
surplus of red and deficit of blue occurs when the red color profile is already raised at a highlight while the green and blue profiles are still low
– an effect that can be caused by chromatic aberration which magnifies
objects in the green and blue image bands. The surplus in red on one side
of the highlight is coupled with a drop of the red profile at the opposite
side while the green and blue profile are still high, resulting in cyan pixels.
Accordingly, the highlight on the green cup is bounded by orange pixels on
one side and cyan pixels on the other side.

   Color Figure 14 (see color insert) shows that the Dichromatic Reflection
Model applies to a variety of dielectric, non-uniform materials. The figure displays the images and the color histograms of a ceramic cup and of
two folders made out of glossy paper under white light. The folders are
slightly bent so that they have highlights in only a small object area. The
corresponding color histograms consist of matte and highlight clusters, as
described above, thus supporting the theory of the Dichromatic Reflection
Model.

3.6 Summary

This chapter has described how camera characteristics influence the measurement of light and its representation as color pixel values in an image.

- Spectral integration, a light filtering operation, reduces the continuous spectrum of a light ray to a finite set of measurements, typically describing it by a red, a green and a blue measurement. Accordingly, the infinite-dimensional vector space of spectral power distributions is projected into a three-dimensional color space.

- The limited dynamic range of cameras limits measurements of incident light to a fixed section of the color space, a color cube. If the incoming light is too bright, its color measurement is clipped along the walls of the color cube. This can also cause color blooming in neighboring pixels. As a result of such camera limitations, the color of pixels in the center of highlights is often not measured according to its physical properties. Such color pixels do not follow the dichromatic theory. In particular, they do not lie in the skewed-T color cluster – and not even in the dichromatic plane.

- Color balancing needs to be performed during the image formation process to account for the changing response of CCD-cameras to incident light at different wavelengths. To obtain equal scaling in the red, green and blue color measurements, we suggest the application of aperture balancing while the images are taken. This method uses different apertures for recording the three image bands.

- A gamma-correction process is built into most commonly available cameras, relating the camera output by an inverse color law to the incident flux. This process introduces curvature into the color space, distorting the linear properties of the Dichromatic Reflection Model. A linearization step is necessary to reverse the gamma-correction process of the camera.

- As a result of chromatic aberration in camera lenses, the blue color band of a color image is slightly magnified when compared to the green band, while the red band is slightly shrunk. Accordingly, color variations in the outer regions of an image are not registered at the same positions in the three color bands. This can have a serious effect on the properties of color variation at highlights in these image regions: The highlight clusters are spread in the color space.

Examples at the end of this chapter demonstrate that the Dichromatic Reflection Model adequately describes color variation on non-uniform, opaque dielectrics. The examples also show the influence of the above camera limitations on the color pixel values and on the color histograms.

4

Color Image Segmentation

This and the following chapter describe how the Dichromatic Reflection Model and the sensor model can be transformed into a computational procedure for analyzing color images. This chapter describes how the models can be used for color image segmentation. The goal of segmentation is to identify objects in an image, as delineated by material boundaries. Because most color segmentation methods in the past have considered object color to be a constant property of an object and color variation to be caused by random camera noise, they generally segment images not only along material boundaries but also along other lines exhibiting color variations, such as highlight and shadow boundaries, or object edges with significant shading changes. The Dichromatic Reflection Model provides an interpretation scheme that relates the physics of light reflection to color changes in the image and the color space. The approach to color image segmentation described here uses the Dichromatic Reflection Model to distinguish relevant color changes at material boundaries from irrelevant ones due to shading or highlights.

4.1 Color Image Analysis Guided by the Dichromatic Reflection Model

The previous chapters have analyzed how shading, highlights and camera limitations influence color variation in an image. It is now our goal to invert this line of reasoning and design an algorithm which inspects an image and concludes from the color variations which optical phenomena in the scene or in the camera have caused it. The problem, however, is that the influence of any such process or combination of processes generally results in the same local image feature: color variation. To distinguish between the influences from different processes, the algorithm needs to inspect extended image areas. It needs to accumulate local color variations to determine distinguishing characteristics of the processes, such as the T-shape of a color cluster. But how can the optimal extent of an area be determined? This seems to be a circular "chicken and egg" problem of, on the one hand, needing a prior segmentation to relate color variation to physical processes and, on the other hand, needing an understanding of the physical processes to provide a good segmentation.

4.1.1 Global versus Local Color Image Analysis

Conceptually, there are two different ways to approach this problem. The algorithm may either start out with very large image areas and subsequently shrink or split them until the regions correspond to objects in the scene, or it may start with small local areas and then merge or grow them.

Within the framework of the Dichromatic Reflection Model, the first ("global") method has to distinguish material boundaries from physical effects in the scene by globally projecting the entire image into the color space and then applying analysis techniques to the color space to identify skewed T's in the color space. It determines material boundaries by distinguishing between several skewed T's. and encounters problems when several different objects with very similar colors existed in the scene, in that the color clusters from different objects would overlap in the color space. The severity of this problem can be observed in Color Figure 15 (see color insert). The figure displays the color histogram of the image in Color Figure 1, showing a scene of eight plastic objects under white light. The various clusters overlap significantly.

The second ("local") method starts out with small image areas and merges them or grows them into neighboring areas that fit the same skewed T. Since local color variation may be relatively small on flat or dark objects, it may be similar in magnitude to camera noise. For this reason, the local approach has to account for the problem of distinguishing camera noise from systematic color variation. It might do this by relying on the assumption that it can divide the image into small, basic image areas that are large enough to indicate the systematic color variation of highlights and shading, but small enough that most of them lie within one object.

4.1.2 Overview of the Approach

This monograph uses the local approach to color image understanding. To determine the extent of local image areas that fit the same skewed T, the approach alternates between generating physical hypotheses from local image data and verifying whether the hypotheses fit the image. The data flow is shown in Figure 4.1. The hypotheses relate object color, shading, highlights and camera limitations to the shapes of color clusters in local image areas. The algorithm searches in a bottom-up manner for color clusters from local image areas that exhibit the characteristic features of the body and/or surface reflection processes. When it finds a promising cluster in an image area, it generates a hypothesis that describes the object color and/or highlight color in the image area and determines the shading and highlight components of every pixel in the area. It then applies the new hypothesis to the image, using a region-growing approach to determine the precise extent of the image area to which the hypothesis applies. This step verifies the applicability of the hypothesis. The physical knowledge embedded in the new hypothesis is used to split every pixel in the new image area into its intrinsic body and surface reflection components. The resulting intrinsic images and the hypotheses together instantiate the general concepts of shading and highlights of the Dichromatic Reflection Model, describing the specific reflection processes that occur in this part of the scene.

The above interpretation scheme can be used to incrementally generate physical information about the scene. Starting out with the simplest and most common aspects of light reflection, an image analysis system can build on the established knowledge and adapt its behavior to local scene properties to address more complicated sections. Accordingly, the image analysis system described here performs its generate-and-test analysis in several stages, each of which is related to a particular aspect of the Dichromatic Reflection Model or the sensor model, as shown in Figure 4.2. It starts out by looking for large matte color clusters, generating hypotheses on the positions and orientations of matte clusters in the color space, as well as determining their spatial extent in the image. For this step, it uses an initial, rough description of local color variation in small image windows. To account for surface reflection on the objects, it then extends the hypotheses into descriptions of skewed T's by combining matte object areas with highlight areas. It analyzes the effects of blooming and color

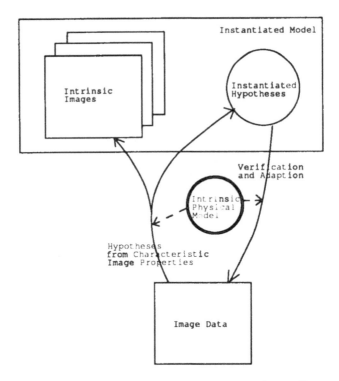

Figure 4.1. Using an intrinsic model for image understanding.

clipping in highlight areas and, finally, it exploits the hypothesized shape of the skewed T's to split the color pixels into their reflection components, generating intrinsic reflection images of the scene. Each of these steps will be described in detail below.

The control structure described above exploits a physical model of light reflection to incrementally identify local and global properties of the scene, such as object and illumination colors. It uses these properties to interpret the pixels in an image. By using this control structure, the image interpretation process can be adapted to local scene characteristics so that the system reacts differently to color and intensity changes at different places in the image. This differs from work done by Gershon [45], which begins with a traditional segmentation method followed by a physics-based postprocessing step. Gershon's method suffers from erroneous region boundaries created by the initial segmentation.

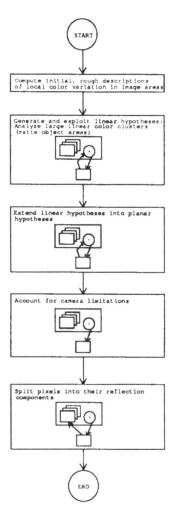

Figure 4.2. Interpretation stages.

4.2 Generating Initial Estimates for Color Clusters

To start the image analysis process, some rough initial estimates about the color variations in the image are necessary. These initial estimates may be too coarse to be useful as hypotheses without further considerations, but they can form the basis for the formulation of hypotheses about linear and planar color variation.

4.2.1 A Classification Scheme for Local Color Variation

To estimate and classify local color variation in the image, the segmentation system described here divides the image[1] into small, non-overlapping windows of fixed size[2]. It projects the color pixels from one window at a time into the color space and finds the principal components of the color distributions, as indicated by the eigenvectors and eigenvalues of the covariance matrix of the cluster [4, 51]. The eigenvalues and eigenvectors determine the orientation and extent of the ellipsoid that optimally fits the data. For convenience, the eigenvalues are sorted by decreasing size, such that the first eigenvalue is the largest one, the second eigenvalue the medium one and the third eigenvalue the smallest one: $\lambda_1 \geq \lambda_2 \geq \lambda_3$.

The shape of the ellipsoids provides information which relates local color variation to physical interpretations. The classification is based on the number of eigenvalues that are approximately zero, within the limit of pixel value noise, σ_0, in the image. The value σ_0 is determined by the estimated amount of camera noise[3]. The decision for each eigenvalue is based on a χ^2-test with $(n^2 - 1)$ parameters, where n is the window size[4]. Each color cluster is then classified according to how many eigenvalues are significantly greater than zero. Figure 4.3 illustrates how the initial windows are classified. A summary of the classification is given in Table 4.1.

In zero-dimensional (*pointlike*) clusters, all three eigenvalues of the window are very small. No significant color variation exists in such a window, and the body and surface reflection components are nearly constant. This is the case if the window lies on a very flat object, such that the photometric angles are nearly constant within the window. Furthermore, dark objects show little variation in shading, independent of their surface curvature. In the extreme, a perfectly black object does not exhibit any shading variation. If a window is selected on a dark object outside a highlight area, it generates a pointlike cluster.

One-dimensional (*linear*) clusters are clusters for which only the first eigenvalue is significantly larger than the estimated camera noise. Pixels in such a window may come from a matte object area forming part of a matte cluster. They also come from the interior of a highlight area and form part of a highlight cluster. As a third possibility, the window may overlap the matte object areas of two neighboring objects that are very dark or flat. Such a window consists of two pointlike color clusters or of

---

[1] A typical image size is 480 × 512 pixels.

[2] Typically 10 × 10 pixels.

[3] For the images shown here, the estimated camera noise was experimentally determined to be about 2.5.

[4] The significance level is $\alpha_0 = 0.005$.

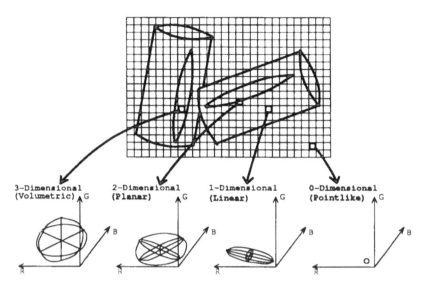

Figure 4.3. Color classes: (a) three-dimensional (volumetric); (b) two-dimensional (planar); (c) one-dimensional (linear); (d) zero-dimensional (pointlike).

one dark pointlike cluster and one linear cluster which together fit well into one linear ellipsoid. A combination of a dark pointlike cluster and a linear cluster of neighboring matte object areas fits a linear ellipsoid, because all matte clusters converge at dark pixels and the dark pointlike cluster lies very close to the matte line defined by the linear cluster. Such cases arise, for example, if a curved object partially occludes a matte part of a neighboring object and the phase angle between viewing and illumination direction is small. At such an occluding boundary, the occluding object is very dark, because the surface normal is nearly perpendicular to the viewing direction and illumination direction. The color on the occluded object, on the other hand, can be arbitrarily bright and may exhibit large color variation.

Two-dimensional (*planar*) clusters have large first and second eigenvalues. The local color data fits a plane in the color cube. Such clusters occur at windows that cover some matte and some highlight pixels of one object. In this case, the third eigenvector of the color cluster determines the normal to the dichromatic plane of the object. Planar clusters also arise, however, in windows that overlie matte pixels from two neighboring regions. Such windows have planar clusters because the matte clusters of

Table 4.1. Summary of color cluster interpretations.

$\lambda_1\ \lambda_2\ \lambda_3$	class		interpretation
s s s	point	a)	window on one object with little curvature
		b)	window on one object with dark color, not on highlight
l s s	line	a)	window on one object: matte cluster
		b)	window on one object: highlight cluster
		c)	window on two objects: matte clusters
l l s	plane	a)	window on one object: dichromatic plane
		b)	window on two objects: matte clusters or highlight clusters
l l l	volume	a)	window on one object: noise (clipping or blooming)
		b)	window on two objects: both matte and highlight pixels on one object, matte and/or highlight pixels on the other object
		c)	window on more than two objects

with s = small and l = large

all objects converge at dark pixels, yielding matte clusters that are pairwise coplanar.

In three-dimensional (*volumetric*) clusters, all three eigenvalues are large. Such color clusters may arise in the middle of highlights where color clipping and blooming significantly increase the noise in the pixel measurements. Volumetric color clusters also occur along material boundaries when three or more objects of different colors share a window or when a window overlies matte pixels of one object and matte and highlight pixels of another object.

Color Figure 16 (see color insert) shows the classification of the color clusters from initial 10×10 windows in the image of the eight plastic objects (Color Figure 1). Pointlike clusters are displayed in yellow, linear clusters in red, planar clusters in green, and volumetric clusters in blue. The image shows that the classifications relate in the expected way to scene properties. Most matte object areas are covered by linear windows, while windows at material boundaries and at highlights are planar or volumetric.

4.2.2 Determining Compatible Color Variation

If the window size is chosen appropriately, it is small compared to the sizes of the objects in the scene. Most windows then lie inside object areas, and only few windows overlie material boundaries. There will generally be many windows within a single object area. All linear clusters of such windows indicate the direction of the same matte or highlight cluster, and all planar clusters support the same planar hypothesis about a dichromatic plane. This section describes how such windows can be merged to provide increased support for a small number of estimates about color clusters in the image.

The algorithm merges neighboring windows that have similar color characteristics. It proceeds in row-major order, testing for all pairwise combinations of neighboring areas whether the areas can be merged. In order not to merge windows across material boundaries, it only merges neighboring windows if both of them, as well as the resulting larger window, have the same classification. It uses the χ^2-test described above to classify the larger window. Accordingly, it combines windows with pointlike clusters into larger pointlike windows; it merges windows with linear characteristics into larger linear windows; and it merges planar windows into larger planar windows. It does not merge neighboring volumetric color clusters since there is no constraint on the resulting cluster. This process continues until no more areas can be merged. The results are initial hypotheses about the positions and orientations of pointlike, linear, and planar clusters in color space and their respective approximate extents in the image.

Color Figure 17 (see color insert) presents the results of merging neighboring windows of the same class in the image of the eight plastic objects (Color Figure 1). The image provides a few large image areas that are related to matte object areas. Due to the coarse window size, such matte image areas correspond only roughly to the actual matte object areas. They are generally surrounded by many smaller areas close to material boundaries or highlights. Only very few planar windows in highlight areas have been merged, probably because of the spread in the highlight clusters due to camera limitations.

4.3   Generating Linear Hypotheses

The initial estimates of color clusters can be used to generate hypotheses about the existing color variation in the image in the subsequent steps. The above discussion and Color Figure 17 show that the initial estimates are rather coarse indications of local color variation. There are far more image areas than objects and, due to the coarse initial window width, the area boundaries do not coincide with the material boundaries. Neighboring areas and their estimates on local color variation now need to be combined at a much finer granularity, resegmenting the image by considering each pixel on an individual basis. The segmentation system selects suitable initial estimates of local color variation and uses them to resegment the image. The new estimates are the basis for hypotheses which relate the color clusters to properties of the Dichromatic Reflection Model, describing color pixels as matte or highlight pixels or as pixels showing camera problems.

The algorithm starts by choosing large image areas with linear color clusters as its hypotheses. Large linear clusters generally correspond to the internal windows from large matte object areas. Such windows are less influenced by estimation errors in local color variation than small image areas which often exist along material boundaries and in highlights where camera limitations or neighboring objects spread the color clusters. Large linear color clusters are also more suitable than planar, pointlike or volumetric color clusters, because their first eigenvector describes the orientation of major color variation in the cluster - information which can be related to a physical interpretation. The eigenvectors of planar clusters (e.g., in areas close to highlights) are less useful because the directions of the first and second eigenvectors are linear combinations of the directions of several linear clusters (e.g., of a matte and a highlight cluster). Their direction does not describe the branches of the skewed T of a dichromatic cluster, and they cannot be related to the physical properties of body and surface reflection.

4.4   Exploiting Linear Hypotheses

So far, the image analysis has proceeded in a bottom-up manner, extracting information about the scene from the image. In a top-down step, a selected linear hypothesis can now be used to resegment the image locally. The hypothesis provides a model of what color variation to expect on the object part comprising the image area. The mean value and the first eigenvector describe the position and orientation of a linear color cluster, while the second and third eigenvalues determine the extent of the color cluster perpendicular to the major direction of variation. According to the Dichromatic Reflection Model, color variation along the major axis can be attributed to a physical property of the scene, e.g., a changing amount of body or surface reflection or a material boundary. Color variation perpendicular to the first eigenvector is attributed to random noise. Accordingly, linear color clusters can be modeled as cylinders (see Figure 4.4). The axes and positions of the cylinders in the color space are determined by the first eigenvectors and the mean color values of the color clusters. The radius depends on the estimated camera noise and is generally a constant

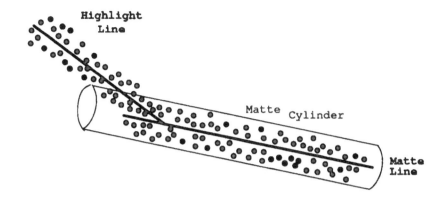

Figure 4.4. Linear hypothesis.

multiple of σ_0[5]. Since all matte clusters merge near the dark corner of the color cube, and dark pixels thus have ambiguous colors, such pixels need to be excluded from the color cylinders. According to such a *dark heuristic*, the cylinders have to be bounded at their dark end by a sphere which is centered at the dark corner of the color cube[6].

The segmentation system can use the color cylinder of the current hypothesis for a linear region growing approach. It selects a start pixel from the image area associated with the color cluster. This pixel must have a color that is contained within the color cylinder. The algorithm then grows a four-connected region from this starting point, recursively examining the four neighbors of pixels on the fringe of the region, and including them if their color lies within the color cylinder. The result is an image area of pixels that are consistent with the current linear hypothesis. The boundary of this new region may be very different from the image area that was initially associated with the linear hypothesis. Due to the coarse initial window size, the initial image area may have contained pixels from neighboring objects or highlight pixels with some amount of surface reflection. Such pixels are excluded from the new region. The initial window may also have contained dark pixels in very shaded object areas or – as is the case in the plastic scene – pixels from a dark background. Due to the dark heuristic, such pixels are also excluded from the new region. On the other hand, the initial segmentation may have had several neighboring image areas on the same matte object part that could not be merged because the characteristics

---

[5] Typically $4\sigma_0$.
[6] A typical radius for the sphere is 23.

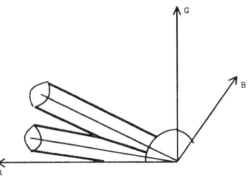

Figure 4.5. Proximity heuristic to resolve a color conflict between the color clusters of two neighboring matte object areas.

of the color clusters were biased by pixels that came from other objects. Many pixels in such neighboring image areas may lie on the current matte object part and be consistent with the current linear hypothesis. They are included into the new segmentation. As a consequence, the new region can be significantly larger than the initial image area, while neighboring image areas may have shrunk.

Since there is generally more than one object in a scene, the system iterates the above steps for each large image area with a linear cluster, selecting the areas by decreasing size. It stops when the next selected area is too small[7]. Since all matte clusters converge at dark pixels, there exists a potential conflict between neighboring matte areas. The dark heuristic eliminates the most difficult cases, although the cylinders of neighboring clusters may still intersect beyond the selected dark threshold. This depends on the cylinder radius and on the angle between the two cylinders. Neighboring objects with very similar colors have conflicts even at fairly bright colors. Pixels with a color conflict are assigned to the cluster with the closest axis, as shown in Figure 4.5. This is called the *proximity heuristic*.

In principle, the linear regions resulting from the use of this system may be related in any of several different ways to the physical processes in the scene. As discussed in Section 4.2.1, a linear color cluster may be a matte cluster, a highlight cluster or a combination of two clusters across a material boundary. Linear color clusters from highlights and across material boundaries are generally much smaller than matte clusters. Since this section considers only linear hypotheses of large image clusters, the

---

[7]Typically less than 500 pixels.

following sections assume that all newly grown regions correspond to matte linear clusters.

Color Figure 18 (see color insert) shows the results of selecting and applying linear hypotheses to the image with the eight plastic objects. The region boundaries outline the matte object parts in the scene, with the material boundaries being well observed. The highlight areas on the objects have not yet been analyzed. This sometimes divides the matte pixels on an object into several matte areas, as shown on the green donut in the lower right part of the image.

4.5   Generating Planar Hypotheses

The Dichromatic Reflection Model states that pixels from a single object lie in a plane in the color space. The linear hypotheses and the linear segmentation generated so far provide a good starting point to describe color variation in an image, but they do not account for the effects of surface reflection. The hypotheses will now be extended into planar hypotheses which describe dichromatic planes and skewed T's. In principle, information on color variation in a dichromatic plane may be obtained from inspecting the characteristics of the initially determined planar color clusters. However, as discussed in Section 4.3, such information is generally not very reliable, due to the small size of most planar image areas. Furthermore, the eigenvectors of such planar clusters do not describe the skewed-T structure of the color cluster. Instead of using planar color information, the segmentation system uses the existing linear hypotheses to describe one branch of the skewed T's. It now determines the orientations and positions of the second branch of the skewed T's.

In this process, the algorithm considers all neighbors of a linear region as prospective highlight candidates. Some neighboring regions may be matte areas on neighboring objects while others may be a part of a highlight on the current object. These regions must be distinguished from one another so that color information from two neighboring matte image areas is not incorrectly combined into a planar hypothesis. To distinguish neighboring matte regions from highlight regions, Gershon assumed that neighboring matte clusters are parallel to one another and thus tested whether two clusters are nearly parallel or whether they intersect [47]. We replace this test by the 50% heuristic of Chapter 2: The color clusters of a matte and a highlight region on the same object intersect in the upper 50% of the matte cluster, while color clusters from neighboring matte regions converge at dark pixels. Accordingly, the algorithm tests whether the two clusters form a skewed T and meet in the upper half of the matte cluster. For this

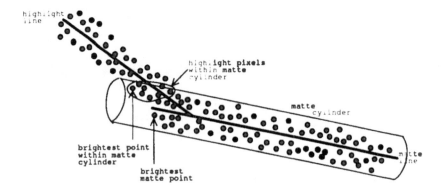

Figure 4.6. Finding the brightest matte pixel.

purpose, it searches for the brightest matte pixel in the color cluster to determine the length of the matte line However, the matte cylinder may contain some highlight points from the foothill of the highlight cluster, as shown in Figure 4.6. Such highlight pixels may be brighter than the brightest matte pixel. They can be distinguished from matte pixels by the observation that highlight clusters always grow inward into the color cube, due to the additive nature of body and surface reflection. The algorithm therefore chooses the brightest matte pixel only from pixels with color values on the outside of the matte line.

Once the length of the matte line is determined, the intersection points of the current matte cluster and of the neighboring prospective highlight cluster can be computed and related to the length of the matte line. The color means and first eigenvectors of the matte and highlight clusters determine the two points at which the respective matte and highlight lines are closest to one another. If the distance between them is larger than a multiple of the estimated camera noise[8], it is unlikely that the clusters meet in a skewed T, and the neighboring area can be discarded as a highlight candidate. Similarly, if the clusters intersect in the lower 50% of the matte cluster, they are probably two matte clusters, and, again, the neighbor can be discarded.

The segmentation algorithm also checks whether the neighboring cluster has a positive direction, a test related to the additive properties of body and surface reflection. If this is not the case, the window of the highlight candidate probably overlies a material boundary, partially covering pixels from the current matte region just outside its color cylinder and partially

[8]Typically $4\sigma_0$.

covering pixels from a neighboring matte region. Such neighboring image areas are discarded from the list of prospective highlight candidates.

After all these tests, there may still exist several highlight candidates because the highlight on the object may consist of a series of small windows that could not be merged, due to color clipping and blooming in the middle of the highlight. Since many such highlight areas are very small, and since they may contain some clipped and bloomed pixels, the orientations of their color clusters may vary significantly. To select a good representative of the entire highlight, the algorithm averages the intersection points of all highlight candidates, weighted by the number of pixels in the regions. It then selects the highlight region whose intersection point is closest to the average.

Since the surface reflection color of dielectrics is generally very similar to the illumination color, all highlight clusters are parallel to one another and to the illumination color vector. As a consequence, all dichromatic planes intersect along one line, which is the illumination vector. This constraint can be used to reduce the error in estimating the orientations of the highlight clusters further. Accordingly, the segmentation algorithm computes the average direction of all highlight clusters, combining all highlight hypotheses into a single hypothesis on the illumination color vector.

The existing linear hypotheses can then be extended into planar hypotheses by combining them with the direction of the illumination vector as well as with information on the starting points of the respective highlight clusters. Such planar hypotheses describe the skewed-T shapes of the color clusters in terms of the positions and orientations of the matte and highlight clusters. The cross product of the illumination vector and the first eigenvector of a matte cluster determine the normal to the dichromatic plane of a color cluster. The position of the dichromatic plane in the color space is given by the color mean.

4.6   Exploiting Planar Hypotheses

Starting from the already generated linear segmentation, the image can be resegmented, using the planar hypotheses to account for the effects of surface reflection on objects. Applying the planar hypotheses one at a time, the algorithm proceeds iteratively until no more unprocessed planar hypotheses for large image areas[9] exist. The chosen planar hypothesis describes the position and orientation of a dichromatic plane in the color space, as well as the shape of the color cluster within the plane. To account

---

[9]Typically containing at least 500 pixels.

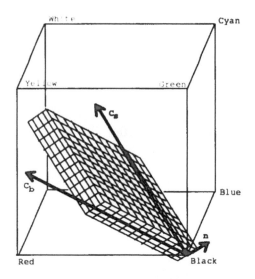

Figure 4.7. A planar slice.

for camera noise, the plane can be extended into a slice of fixed thickness[10], as shown in Figure 4.7.

When the segmentation algorithm uses the chosen planar slice to re-segment the image locally, it starts, in principle, from the selected matte region and expands it until no more pixels at the region boundaries fall into the planar slice. This planar region growing method is augmented with special provisions to handle coplanar color clusters from neighboring objects. Such coplanar color clusters occur when the illumination vector lies in the plane spanned by the matte vectors of two neighboring objects. The color clusters of such objects lie in the same dichromatic plane and cannot be distinguished in a simple planar region growing approach. The resulting segmentation would generally be quite counterintuitive, since the colors of objects with coplanar clusters may be very different and even complementary. For example, a red body reflection vector $(255, 0, 0)$ and a white illumination vector $(255, 255, 255)$ span a dichromatic plane with normal $(0, -1/\sqrt{2}, 1/\sqrt{2})$. This plane contains all body reflection vectors (r, g, b), for which the g-component and the b-component are identical. In particular, the plane contains the vector $(0, 255, 255)$, which describes a cyan object color, as shown in Figure 4.8. In this case, a simple region

---

[10] A typical choice for the width of the slice is $4\sigma_0$ in the positive and negative direction of the normal.

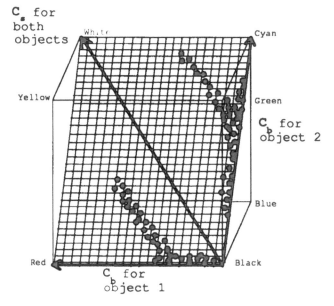

Figure 4.8. Coplanar color clusters.

growing method would not be able to distinguish a red object from a cyan object.

Such segmentation problems can be avoided if the previously gathered knowledge about existing matte color clusters is included into this segmentation phase: When the planar region growing process encounters pixels from a previously grown matte region other than the starting region, it should only continue growing if the pixel lies within the matte color cylinder of the starting region. Accordingly, the unrestricted planar growing criterion is applied only to pixels that have not been previously recognized as matte pixels, while the linear region growing method is used when matte pixels are concerned. This reflects the observation that if several matte areas exist in an object area, separated by highlight regions, all such matte areas form a single matte cluster. The segmentation algorithm uses this strategy. It also applies the proximity heuristic described in Section 4.4 to resolve ambiguities for color pixels at the intersection of dichromatic planes.

Color Figure 19 (see color insert) displays the results of segmenting the scene using the generated planar hypotheses. In comparison to the linear segmentation in Color Figure 18, the segmented image areas have grown ·

into the highlight areas. As a result, two matte image areas on the green donut – previously separated by a highlight – are now united. Due to camera limitations, not all pixels in the centers of the highlights are yet integrated into the object areas. This will be discussed and remedied in the next section.

4.7   Accounting for Camera Limitations

Unfortunately, real images generally do not fully comply with the Dichromatic Reflection Model. Among other things, color clipping and blooming may significantly distort the color of pixels in and around highlights. As a result, the color pixels in the centers of highlights generally do not fall into the planar slice defined for the dichromatic plane of an object area. The planar segmentation excludes such pixels, as can be observed in Color Figure 19.

Since color information is so unreliable for these pixels, it cannot be used to analyze the pixels. It is more appropriate to use geometric considerations instead to include distorted pixels into the region. Starting from highlight pixels, therefore, the algorithm expands the planar regions into areas that are next to highlight pixels and contain very bright pixels (i.e., brighter than the intersection point between the matte and highlight cluster).

Color Figure 20 (see color insert) displays the final results of segmenting the scene while using the generated planar hypotheses and accounting for camera limitations. Nearly all pixels in the highlight centers have now been integrated into the segmented regions, and the image segments correspond quite well to the objects in the scene. A few pixels around the highlights have been excluded, due to the heuristic of only integrating very bright pixels. Any image processing method for filling holes in image segments, such as a region expansion operation followed by region shrinking [156], should be able to include these pixels.

4.8   Optical Effects Beyond the Scope of the Segmentation Algorithm and the Reflection Model

Color Figure 20 shows that some pixels at object borders have not been included into the object areas. These exclusions are related to effects in the scene that are not modeled in the Dichromatic Reflection Model, such as shadow casting and interreflection between objects. It is a strength rather than a weakness of the algorithm that it excludes such pixels from the segmentation. This shows that the algorithm does not impose an incorrect, oversimplified interpretation on pixels that need a more sophisticated explanation. Future work will be involved in modeling these physical effects and in including their analysis into the algorithm.

The segmentation of the yellow donut in the plastic scene exhibits two interesting properties. A small area in the lower right part of the yellow donut was not included into the large image segment covering the donut. The color of these pixels has been significantly altered by interreflection from the orange cup, which reflected orange light onto the yellow donut, altering its body color. Secondly, a small area at the upper left edge of the same donut has wrongly been assigned to the red donut above it. In the original image, this area is covered by a shadow, resulting in very dark pixel values. Since the color clusters of the two donuts (yellow and red) are already merged at these color values, the assignment was based on the distance from the two cylinder axes, which happened to be smaller for the red donut. Interreflection may also have influenced the color values, biasing them towards the red color vector.

Interreflection also occurs between the orange and yellow cup. There is a mirror image of the yellow cup on the side of the orange cup, resulting in yellow surface reflection. Such yellow surface reflection is added to the orange body reflection – which, in addition, has been influenced by the light reflected from the yellow cup. As a consequence, the segmentation did not include these pixels into the region representing the orange cup. Similarly, interreflection between the orange and green cup causes a few pixels on the uppermost planar surface of the polyhedral part of the green cup to be excluded from the region representing the green cup.

Finally, there is a mirror image of the yellow donut on the lower side of the blue donut. This is displayed in Color Figures 21 and 22 (see color insert) which show a magnified part of the blue donut and its color histogram. These pixels on the blue donut have a more yellowish color than the rest of the donut. In the color space in Color Figure 22, this color change appears in the form of an extra branch that extends from the matte cluster toward yellow. Color Figure 18 shows that these yellow pixels were separated from the blue ones in the linear segmentation step, due to their distance from the matte cylinder of the blue donut. They were included into the blue object area during the planar segmentation step, however, because the yellow interreflection cluster was close enough to the dichromatic plane spanned by the blue matte vector and the white illumination vector. This case of interreflection was too subtle to be detected by the algorithm under the chosen set of control parameters – especially under the chosen width of the planar slice. An analysis of the global color cluster shape may be helpful in detecting the interreflection branch of the color cluster and in explicitly excluding it from the planar region growing step (using the same technique used to exclude neighboring coplanar matte clusters).

4.9 Summary

This chapter has described a color image segmentation algorithm that uses a physical model of light reflection. Instead of exclusively attributing color variation to noise and material changes, as traditional statistical signal-based segmentation methods do, this approach accounts for color variations due to shading and highlights. It is guided by the Dichromatic Reflection Model to generate physical hypotheses which relate object color, shading, highlights and camera limitations to the shapes of color clusters in local image areas. The algorithm starts out with a rough initial estimate of local color variation in local image windows, as described by the principle components of the color clusters in the windows. The subsequent three steps each use a generate-and-test approach, generating hypotheses about how to interpret local color variation and applying such hypotheses to the image. The algorithm first looks for linear color variation, relating it to matte object areas. It then expands the matte areas into areas with planar color variation to account for surface reflection on objects. Finally, it accounts for highlight pixels with camera problems. Examples show how these segmentation steps perform on a real image. The final segmentation image demonstrates that the algorithm is able to determine material edges in a scene of eight plastic objects, ignoring color changes due to shading and highlights. Some optical effects in the scene, such as shadow casting and interreflection, show the limited scope of the current model. The algorithm generally excludes areas underlying such effects from its analysis, not imposing an incorrect, oversimplified interpretation on such pixels.

5

Separating Pixels into Their Reflection Components

This chapter shows how the Dichromatic Reflection Model and the segmentation results of Chapter 4 can be used to generate intrinsic body and surface reflection images of the scene. The body reflection image shows the scene without highlights, while the surface reflection image shows only the highlights. Since the reflection images have a simpler relationship to the illumination geometry than the original image, they may improve the results of many other computer vision algorithms, such as motion analysis, stereo vision, and shape from shading or highlights.

The Dichromatic Reflection Model describes the color of every pixel in a color image as a mixture of a body reflection component and a surface reflection component. The reflection images are generated by reversing this color mixing process, exploiting the skewed-T shape of the color clusters. The following section presents and compares two methods of determining the body and surface reflection vectors of a previously segmented image area. One of these methods exploits the knowledge of body and surface reflections that the method gathered while segmenting the image. The other method starts over, using only the produced segmentation. It obtains the surface and body reflection vectors from the shape of entire color clusters or image areas. We discuss strengths and limitations of both methods.

Subsequent sections describe the basic method of splitting color pixels into their body and surface reflection components and how the method can be applied to clipped or bloomed pixels that do not satisfy the assumptions of the Dichromatic Reflection Model.

Note that it is impossible to determine the exact lengths of the reflection vectors from the clusters in color space, since the length of a cluster depends both on the magnitude of the reflection curve of the material and on the geometric influence from the photometric angles: A dark, but strongly curved, object may produce a color cluster of the same length as a brighter, but relatively flat, object. As a consequence, only the orientation of the reflection vectors can be determined from color images, not their absolute lengths. Accordingly, the intrinsic body and surface reflection components of color pixels can be computed only up to a scale factor that depends on the brightness of the material.

5.1   Determining Body and Surface Reflection Vectors

To be able to split color pixels into their body and surface reflection components, it is essential that the body and surface reflection vectors, C_b and C_s, of the pixel be known. The geometric scale factors, m_b and m_s, can then be determined, as will be shown in Section 5.2. The image analysis system described here has two different ways of obtaining the reflection vectors. First, it can use the linear and planar hypotheses about the body and surface reflection vectors that were generated during the segmentation phase. Second, it can determine the reflection vectors from a global analysis of the skewed-T shape of a color cluster generated by projecting all color pixels from a region into the color space. The system has the capability of using either of these methods, at the user's choice. Since both exhibit strengths and limitations in their performance, this section discusses and compares both of them.

5.1.1   Determining the Reflection Vectors from the
Global Cluster Shape

The global algorithm projects the color pixels from a chosen image area into the color cube, ignoring clipped or bloomed pixels as described in Section 3.2. Since the Dichromatic Reflection Model assumes that the color pixels lie in a dichromatic plane in the color cube, the algorithm fits a plane to the cluster, using principal component analysis, as described in Section 4.2.1. The third eigenvector determines the direction of minimal color variation in the cluster. The Dichromatic Reflection Model interprets

color variation along this vector as caused by random camera noise, while
all the physically relevant color variation is expected to occur in the dichromatic plane perpendicular to the vector. Accordingly, the global algorithm
projects all color pixels along the third eigenvector into the dichromatic
plane, thus reducing the camera noise.

The analysis of the relationship between object shape and cluster shape
in Section 2.3 states that color clusters form skewed T's in the dichromatic
plane and that the orientations of the linear sections of the T determine
the body and surface reflection vectors of the cluster. The global algorithm
determines the reflection vectors by examining the shape of the entire color
cluster of a region. It separates the color cluster into two linear subclusters that contain the matte and highlight pixels, respectively. Determining
these linear subclusters in the color space is not trivial, because neither a
classification of the pixels into matte and highlight pixels nor the orientation and position of the subclusters are known. The algorithm determines
the two subclusters in a two-stage process.

The first stage results in rough initial estimates of the linear matte and
highlight subclusters. The algorithm fits a convex hull of line segments to
the contour of the cluster, using a recursive line spitting method [150]. To
do so, it determines the line L between the brightest and darkest points in
the cluster to be the starting line. It divides the plane into buckets which
have the form of slim stripes running perpendicularly to L (see Figure
5.1). It computes the color distance of all pixels from L and stores the
maximal distance within every stripe s as a function $d(s)$. This function
approximates the contour of the color cluster. Its shape, however, is very
susceptible to noise. To concentrate on the large scale color variations,
the algorithm smoothes $d(s)$, averaging its values over a window. It then
approximates the smoothed contour by a convex polygon by determining
the stripe s_i with maximal distance $d(s_i)$ from L. If $d(s_i)$ exceeds a given
threshold, L is split into two segments, and the distances of the color pixels
from the new line segments are recomputed. Both segments are then split
recursively into smaller segments until all distance measurements are below
the threshold. The choice of the threshold determines how significant an
orientation change of the contour needs to be before it influences the shape
of the polygon.[1]

The algorithm next classifies the line segments. The discussion of the
cluster shapes in Section 2.3 states that matte pixels depend only on the
body reflection component, forming a matte linear cluster that approaches
the origin of the color space, that is, the black corner of the color cube.
Accordingly, the algorithm assumes that the line segment closest to the

---

[1] For all results shown here, the $d(s_i)$ was required to exceed 20.

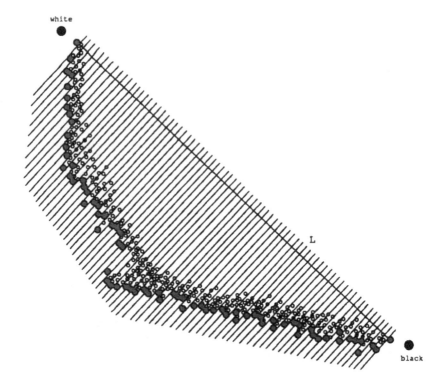

Figure 5.1. Fitting a convex polygon to the color cluster.

black corner of the cube approximates the matte line. Section 2.3 further states that the highlight cluster starts from the matte cluster. Accordingly, the algorithm classifies the line segment connected to the matte line as the highlight line. If there are further line segments connected to the highlight line, the algorithm assumes that they are related to clipped color pixels that were not detected by the heuristics of Section 3.2. All further line segments are thus classified as clipped color lines.

The line fitting method described so far provides an initial estimate for the matte and highlight lines, but, for the following reasons, it cannot be expected to generate very accurate results: Most importantly, the fitted convex polygon does not account for the concavity at the corner between the matte and highlight clusters. As a result, the estimated highlight line generally describes an orientation with too little slope. Secondly, the initial plane fitting step was applied without distinction to all pixels in the color cluster. If some clipped color pixels were not discovered by the discard-

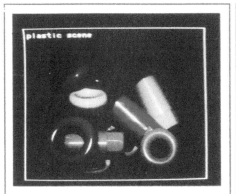

Color Figure 1. Scene with eight plastic objects.

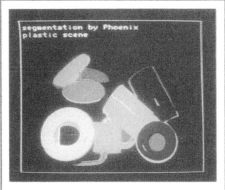

Color Figure 2. Region segmentation by Phoenix, scene with eight plastic objects.

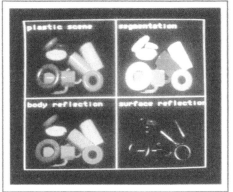

Color Figure 3. Segmentation and intrinsic images using Dichromatic Reflection Model, scene with eight plastic objects.

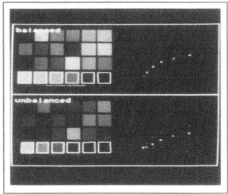

Color Figure 4. The effect of color balancing and gamma correction on color images.

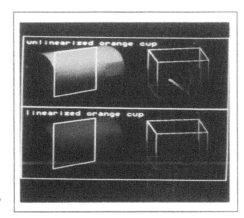

Color Figure 5. Curvature in the color space, caused by gamma-corrected cameras.

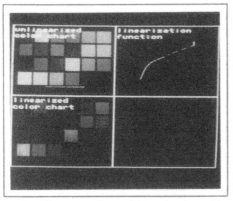

Color Figure 6. Spectral linearization with cubic spline functions.

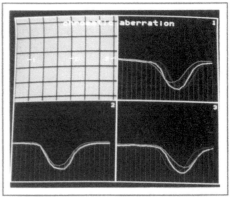

Color Figure 7. Chromatic aberration measurements on a calibration grid.

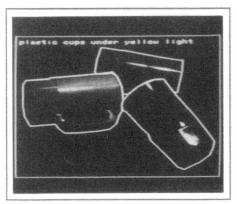

Color Figure 8. Three plastic cups under yellow light.

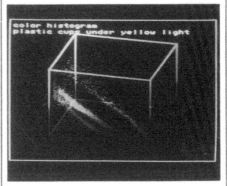

Color Figure 9. Color histogram of three plastic cups under yellow light (clipped colors are shown in blue).

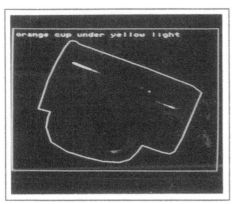

Color Figure 10. Orange cup under yellow light.

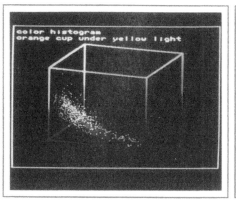

Color Figure 11. Color histogram of the orange cup under yellow light (bloomed colors are shown in white).

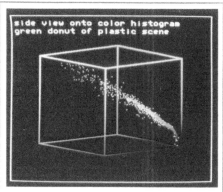

Color Figure 12. Color histogram of the green donut, exhibiting chromatic aberration (side view onto highlight cluster).

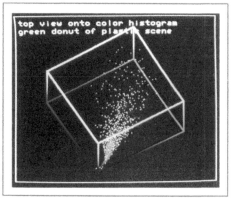

Color Figure 13. Color histogram of the green donut (top view onto highlight cluster).

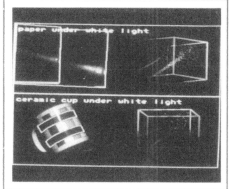

Color Figure 14. Images and color clusters of glossy paper and a ceramic cup under white light.

Color Figure 15. Color histogram of the scene with eight plastic objects.

Color Figure 16. Color cluster classification for initial image areas, scene with eight plastic objects.

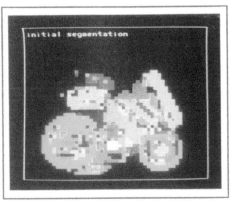

Color Figure 17. Initial grouping into approximate image areas, scene with eight plastic objects.

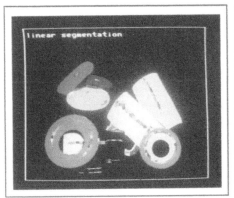

Color Figure 18. Linear segmentation, scene with eight plastic objects.

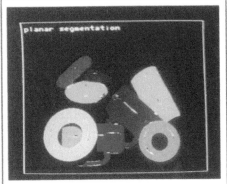

Color Figure 19. Planar segmentation, scene with eight plastic objects.

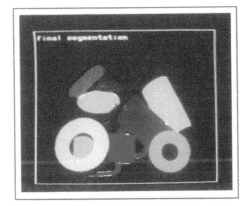

Color Figure 20. Final segmentation, scene with eight plastic objects.

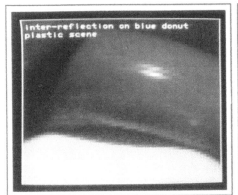

Color Figure 21. Interreflection between the yellow and blue donut, magnified portion of the plastic scene.

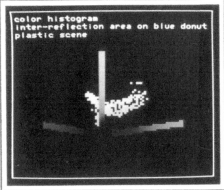

Color Figure 22. Influence of interreflection between the yellow and blue donut onto the color histogram of the blue donut, plastic scene.

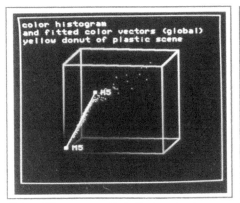

Color Figure 23. Globally fitted color vectors for the yellow donut of the plastic scene.

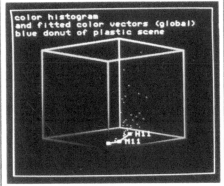

Color Figure 24. Globally fitted color vectors for the blue donut of the plastic scene.

Color Figure 25. Body reflection image, plastic scene (local). (Clipped pixels and bloomed pixels are shown in black).

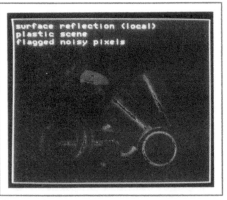

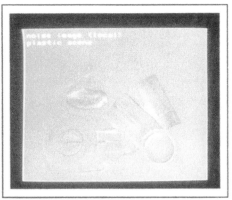

Color Figure 26. Surface reflection image, plastic scene (local). (Clipped and bloomed pixels are shown in black)

Color Figure 27. Noise image, plastic scene (local).

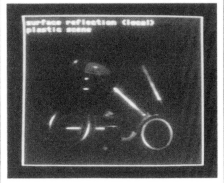

Color Figure 28. Body reflection image, plastic scene (local) (restored clipped and bloomed pixels).

Color Figure 29. Surface reflection image, plastic scene (local) (restored clipped and bloomed pixels).

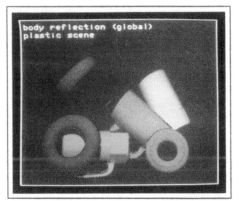

Color Figure 30. Body reflection image, plastic scene (global) (restored clipped and bloomed pixels)

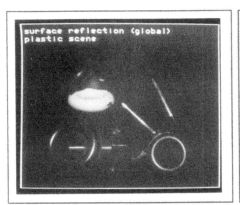

Color Figure 31. Surface reflection image, plastic scene (global) (restored clipped and bloomed pixels)

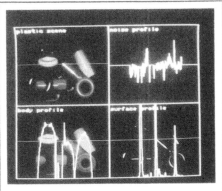

Color Figure 32. Profiles of the reflection images along row 226, plastic scene (local).

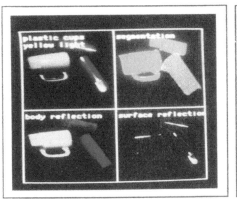

Color Figure 33. Color image segmentation and reflection analysis, scene with three plastic cups under yellow light.

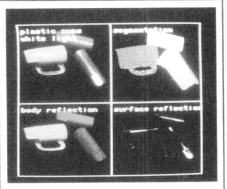

Color Figure 34. Color image segmentation and reflection analysis, scene with three plastic cups under white light.

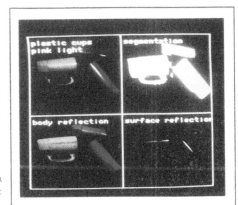

Color Figure 35. Color image segmentation and reflection analysis, scene with three plastic cups under pink light.

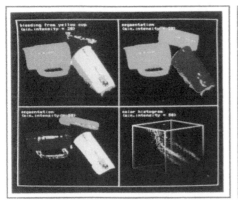

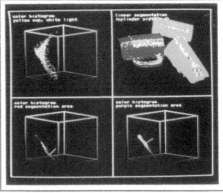

Color Figure 36. The influence of thresholding intensity to exclude dark pixels, three cups under yellow light.

Color Figure 37. The influence of cylinder width, three cups under white light.

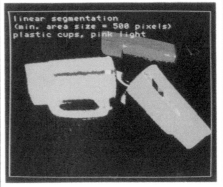

Color Figure 38. The influence of the initial window size, three cups under pink light.

Color Figure 39. The influence of the minimal area size for matte linear clusters, three cups under pink light.

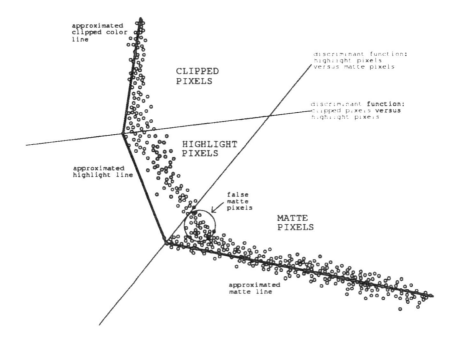

Figure 5.2. Classifying the color pixels.

ing heuristics, they have biased the orientation of the dichromatic plane towards the white corner of the color cube. Since the fitted lines lie in the plane, they are also biased towards white. For these reasons, the algorithm considers the matte and highlight lines to be only an initial approximation of the linear parts of the color cluster. The orientations and positions are improved in the second stage of the line fitting process.

The second stage uses the preliminary result of the first stage to classify the color pixels coarsely into matte pixels, highlight pixels and clipped pixels, depending on the line to which they are closest (see Figure 5.2). To improve the orientations of the matte and highlight lines, the algorithm refits them separately to the matte pixels and to the highlight pixels, using principal component analysis. The result is shown in Figure 5.3. The first eigenvectors describe the directions of the largest color variation in the two clusters. They determine new estimates for the matte and highlight lines with corresponding body and surface reflection vectors parallel.

The above described steps assume that the classification assigns correct labels to most of the pixels in the cluster. Some highlight pixels may be

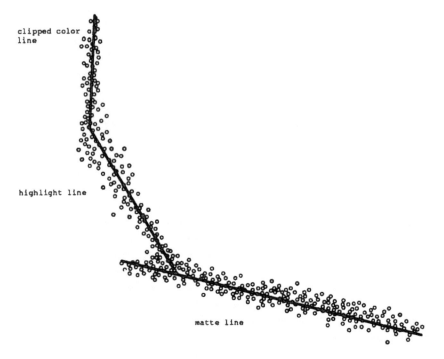

Figure 5.3. Globally fitted reflection vectors.

misclassified as matte pixels because the initial estimate of the highlight
line might have been wrong. Some highlight pixels near the starting point
of the highlight line (i.e., in the concavity) may be closer to the initial
matte line with than to the initial highlight line (see Figure 5.2). Since the
number of such false matte pixels is generally very small, when compared
to the number of correct matte pixels, this will usually not be a serious
problem. If the number of false matte pixels is large, as can be the case
with flat objects when the highlight areas are comparable in size to the
matte areas, the algorithm may fail. It may then have to alternate several
times between reclassifying the pixels and refitting the lines. Such error
recovery mechanisms are an area of future work.

5.1.2 Determining the Reflection Vectors from Local Color Variations

Instead of using global shape analysis to determine the reflection vectors of an image area, the color image analysis system can also exploit

information from local color variation, using the hypotheses provided by the segmentation algorithm. Due to possible estimation errors in the segmentation process and to camera problems, the vectors may not yet fit perfectly. To obtain a more precise fit to the data of a region, all pixels that are not clipped are reclassified as matte or highlight pixels, depending on whether they are closer to the matte line or to the highlight line.[2] The local algorithm then refits the matte and highlight lines to the matte and highlight pixels by determining the first eigenvectors and the color means of the clusters. These lines determine improved directions of the body and surface reflection vectors. Finally, the illumination color vector is recomputed from the improved estimates of the reflection vectors, as described in Section 4.5.

5.1.3 Comparison

The global and local algorithms each have their own strengths and weaknesses. Tables 5.1 and 5.2 show the results of applying both methods to the image of the plastic scene (Color Figure 1), using the segmentation of Chapter 4 (Color Figure 20). At the time of this research, the body reflection vectors can only be evaluated qualitatively since no independent measurement (e.g., by a spectroradiometer) of the body reflection colors is available. The quality of the surface reflection vectors can be evaluated in more quantitative terms: Since the surface reflection component of dielectrics has approximately the same color as the illumination, the surface reflection vectors must be approximately parallel to one another and to the illumination color vector. For evaluation purposes, an independent estimate of the illumination color is obtained by taking a picture of a grey chart[3] under the light illuminating the scene. The chart reflects the illuminating light without changing its spectral curve significantly. The averaged color pixel values of the image serve as an indication of the color of the light source. The resulting, normalized estimate for white tungsten light, measured under aperture balancing by the camera in the Calibrated Imaging Laboratory, is shown in Tables 5.1 and 5.2 as an independent measurement.

For all objects except for the yellow and blue donuts, both the global and local methods appear to provide reasonable estimates for the body and surface reflection vectors of the objects. The estimation differences for body vectors are small. The estimated surface reflection vectors vary to some degree around the illumination vector. The averaged estimates for the illumination color for both methods, however, are quite close to the

---

[2]The orientation and position of the highlight line are determined by the surface reflection vector and the intersection point of the skewed T.

[3]Munsell N 8.5/ with 68.4% reflectivity.

Table 5.1. Body and surface reflection vectors of the eight plastic objects under white light (global).

	body reflection vector	surface reflection vector
dark red donut	(0.99,0.10,0.13)	(0.58,0.57,0.58)
orange cup	(0.95,0.28,0.16)	(0.64,0.50,0.58)
yellow cup	(0.84,0.52,0.14)	(0.48,0.57,0.67)
green cup (right half)	(0.26,0.89,0.37)	(0.51,0.54,0.67)
yellow donut	(-0.89,0.00,0.45)	(0.77,0.61,0.19)
bright red donut	(0.98,0.18,0.11)	(0.70,0.50,0.51)
green donut	(0.18,0.81,0.56)	(0.50,0.61,0.62)
green cup (left half)	(0.25,0.89,0.39)	(0.56,0.52,0.65)
blue donut	(-0.22,0.06,0.97)	(0.17,0.37,0.91)
illumination vector		
- average surf. refl.		(0.56,0.55,0.62)
- independent meas.		(0.58,0.57,0.58)

independent measurement, demonstrating that both methods are suitable estimators for the illumination color. Table 5.3 shows the angles between the independently measured illumination vector and the globally or locally estimated surface reflection vectors, as well as the orientations of the difference vectors. The * indicates which method provides the better estimate for an object.

Table 5.2. Body and surface reflection vectors of the eight plastic objects under white light (local).

	body reflection vector	surface reflection vector
dark red donut	(0.99,0.11,0.13)	(0.63,0.52,0.58)
orange cup	(0.95,0.26,0.14)	(0.68,0.48,0.55)
yellow cup	(0.84,0.52,0.14)	(0.50,0.63,0.59)
green cup (right half)	(0.27,0.89,0.37)	(0.53,0.58,0.62)
yellow donut	(0.77,0.61,0.19)	(0.70,0.50,0.51)
bright red donut	(0.98,0.17,0.11)	(0.67,0.54,0.51)
green donut	(0.19,0.79,0.58)	(0.45,0.64,0.62)
green cup (left half)	(0.27,0.87,0.42)	(0.61,0.54,0.58)
blue donut	(-0.08,0.19,0.98)	(0.48,0.65,0.59)
illumination vector		
- average surf. refl.		(0.59,0.57,0.58)
- independent meas.		(0.58,0.57,0.58)

Table 5.3. Difference vectors and angles between the surface reflection vectors and the independently measured illumination vector of the eight plastic objects under white light.

	global difference vector: angle	local difference vector: angle
dark red donut	(0.00,0.00,0.00): 0.00 *	(0.05,0.05,0.00): 4.05
orange cup	(0.06,0.07,0.00): 5.30 *	(0.10,0.09,0.03): 7.92
yellow cup	(0.10,0.00,0.09): 7.71	(0.08,0.06,0.01): 5.77 *
green cup (right)	(0.07,0.03,0.09): 6.76	(0.05,0.01,0.04): 3.71 *
yellow donut	(0.19,0.04,0.39): 25.18	(0.12,0.07,0.07): 8.93 *
bright red donut	(0.12,0.07,0.07): 8.93	(0.09,0.03,0.07): 6.76 *
green donut	(0.08,0.04,0.04): 5.61 *	(0.09,0.07,0.04): 8.79
green cup (left)	(0.02,0.05,0.07): 5.05	(0.05,0.03,0.00): 2.43 *
blue donut	(0.41,0.20,0.33): 32.79	(0.06,0.08,0.01): 7.37 *

Performance of the Global Method The global method determines negative body reflection vectors for the yellow and blue donuts. This is physically impossible. For the yellow donut, the method suggests that the red body component is negative and the green component is zero. As shown in Color Figure 23 (see color insert), the color cluster of this object has a very sparse highlight cluster due to the small size of the two highlights on the donut. As a result, the blooming heuristic discards nearly all highlight pixels, and the brightest pixel in the cluster is a matte pixel. The method to determine a matte and a highlight line then depends very much on the noise characteristics in this matte cluster . The polygon fitting method selected a break point between the two lines at very dark matte pixel values, just at the start of the cluster, where the matte cluster has a spherical cut-off boundary to exclude dark pixels from the color segmentation. As a result, the body reflection vector lies on the spherical cut-off boundary and points in a negative direction, while the surface reflection vector is fitted to the matte cluster. This leads to a long difference vector and a wide angle between the estimated surface reflection color and the independent measurement of the illumination color, as shown in Tables 5.1 and 5.3.

A similar situation exists for the blue donut. The donut has a negative body reflection vector in Table 5.1, but the negative red component is relatively small and the direction of the body reflection vector describes approximately a blue color vector. The negative orientation of the vector is caused by interreflection from the yellow donut onto the blue donut, as discussed in Section 4.8, which results in an extra branch that extends from the matte blue cluster towards yellow (see Color Figure 22). Since the global method performs a principal component analysis as its final step in estimating the reflection vectors and since the matte cluster of the blue donut is rather short and is also related to a small object area in the scene, the resulting body reflection vector is very sensitive to the influence of interreflection. This shows some of the limitations of using principal component analysis for estimating color variation.

Tables 5.1 and 5.3 indicate that the surface reflection vector of the blue donut has been fitted to the matte cluster instead of to the highlight cluster. As was the case for the yellow donut, the highlight cluster is very sparse. Color Figure 24 (see color insert) shows that only the foothill of the highlight cluster remains after the blooming heuristic is applied. As a result, the polygon fitting procedure does not determine the correct starting point of the highlight cluster, and the surface reflection vector mainly describes the orientation of the matte cluster.

Performance of the Local Method   Since the local method determines color variation by principal component analysis instead of by an investigation of the cluster shape, it is not as sensitive to bloomed pixels as the global method. It does not need to use the blooming heuristic and discard pixels with rare colors. As a result, the local method is able to analyze the highlight clusters of the yellow and blue donut which were nearly entirely discarded by the global method. This leads to much better estimates of the surface reflection vectors, although the estimate for the highlight on the yellow cup still shows some deviation from the illumination vector.

The local method also determines a much better estimate of the body reflection vector of the yellow donut than the global method, because it does not depend on the peculiarities of the boundary shape and it does not use the blooming heuristic. However, it has the same problems with the blue donut as the global method: estimating a negative body reflection vector. Since it uses principal component analysis to determine the reflection vectors, the estimated body reflection vector is biased by the interreflection branch of the matte cluster. With principal component analysis being the major module of the local method, this method is thus severely limited in coping with interreflection or other causes for non-linear matte clusters. The local method needs to be combined with an independent global shape analysis to perform better on such images.

Strengths and Limitations of Both Methods   Although the current global method of the color image analysis system also uses principal component analysis and encounters the problems related to it, it will be discussed here with respect to its strengths and limitations as a method that analyzes the global color cluster shapes. When fitting a polygon to a cluster, the method depends on the lengths of the matte and highlight clusters, as well as on the angle between them. If the angle is very wide, so that the matte and highlight clusters are nearly collinear, the polygon fitting procedure may not be able to detect the slight bend in the cluster where the highlight starts. Whether or not the bend is detected depends on its distance from the line which connects the darkest and brightest points in the cluster. The distance, again, is a function of the lengths of the matte and highlight clusters. Thus, this method performs best on objects with well extended matte and highlight clusters, such as objects with highly saturated colors and a lot of variation in shading. The shorter the matte and highlight clusters are, the more pronounced the angle between them will need to be. If either of the clusters is very short and the angle is not acute enough, the global method fails to determine the correct body and surface reflection vectors. The performance of the global method also depends on the size of the highlight on the object. If the highlight is small and sharp, the resulting highlight cluster does not contain many points, and there may be significant gaps between neighboring points in the highlight cluster. Since the global method discards all color space entries with low frequency counts as bloomed or noisy pixels, it may miss most of such a highlight cluster.

The local method, on the other hand, does not need to discard noisy or bloomed pixels since such pixels have less influence on the principal components of a color cluster (which are used in the local method to estimate the local color variation) than on the shape of its boundary (which is used in the global method). On small highlights, the local method generally performs better than the global method. Furthermore, the local method is not as dependent on the entire length of the matte or highlight cluster, since it determines its estimates from local color variation which in most local areas covers only a small portion of the cluster. The local method in general, however, does not determine the precise location of the start of the highlight cluster as well as the global method. Since it first looks for matte linear clusters, it generally misclassifies a significant number of highlight pixels at the foothill of the highlight cluster as matte pixels because they fall into the matte cylinder. Such pixels may even bias the overall estimate for the orientation of the matte cluster. As a result, only the brighter part of the highlight cluster is generally recognized as a highlight cluster. Since the noise characteristics increase for bright highlight pixels, due to color clipping and blooming, the estimated orientation of the highlight line is not very reliable. The predicted starting point of the highlight cluster thus does not always coincide with the globally visible bend in the cluster. This can be a problem if the matte and highlight clusters are nearly collinear. The local method also does not perform well on color clusters

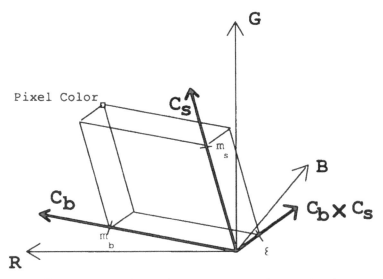

Figure 5.4. Decomposing a color pixel into its constituent body and surface reflection components.

This section describes how $\tilde{m}_s$ and $\tilde{m}_b$ can be determined for every pixel in a region. To split the pixels into the reflection components, the algorithm uses $\overline{\mathbf{C}_b}$, $\overline{\mathbf{C}_s}$ and their cross product, $\overline{\mathbf{C}_b} \times \overline{\mathbf{C}_s}$, to define a new (not necessarily orthogonal) coordinate system in the color cube. This coordinate system describes every color in the cube in terms of their amounts of body reflection, surface reflection and noise ϵ, as given by the color distance from the dichromatic plane (see Figure 5.4). There exists an affine transformation, and thus a linear transformation matrix T, which transforms any color vector $\mathbf{C} = [R, G, B]^T$ from the initial coordinate system into a vector $\mathbf{D} = [\tilde{m}_b, \tilde{m}_s, \epsilon]^T$ in the new coordinate system

$$\mathbf{D} = T\mathbf{C}. \tag{5.5}$$

After computing T from $\overline{\mathbf{C}_b}$ and $\overline{\mathbf{C}_s}$, every color pixel in the region can be transformed into the scaled body and surface reflection components, $\tilde{m}_b$ and $\tilde{m}_s$. The *body reflection image* of the image area consists of the $\tilde{m}_b$-components of all pixels. The corresponding *surface reflection image* consists of the $\tilde{m}_s$-components. The ϵ-components provide the *noise image*.

Color Figures 25 and 26 (see color insert) show the body and surface reflection images that are generated in this way for image of the eight plastic objects under white light (Color Figure 1), using the locally determined reflection vectors from Table 5.2. The colors of clipped and bloomed pixels have not been separated into their reflection components because they do not satisfy the assumptions of the Dichromatic Reflection Model. They need a special analysis which restores the color deficiencies introduced by color clipping or blooming. This analysis will be presented in Section 5.3. In Color Figures 25 and 26, clipped and bloomed pixels are displayed in black, as are the pixels outside the regions corresponding to the objects in the scene. By flagging clipped and bloomed pixels in this way, the algorithm demonstrates that it is able to restrict its analysis to pixels that satisfy the conditions of the dichromatic theory. This information is available in a separate *error image* which may also prove to be a useful source of information to other image understanding algorithms such as shape from shading. It helps them to base their analysis only on undistorted image data. If necessary, algorithms may extrapolate the derived shape information into areas with clipped or bloomed pixels. The error image may also be used to determine the amount of blooming and clipping in an image. When a sequence of images with different apertures is taken, the error image may serve as a criterion to choose an image with little clipping and blooming. Such aspects of the error image have not yet been investigated. They are an area of future work.

Color Figure 25 displays the body reflection image of the plastic scene. Since the body reflection vector of the blue donut has a negative red component, the figure only shows variations in the green and blue components while the red component is set to zero. The body reflection image demonstrates that the approach to color image understanding described here can successfully remove highlights in undistorted image areas. It also shows that the algorithm detects very well the areas in which the pixel values are distorted.

The surface reflection image in Color Figure 26 shows that the system is able to detect highlights in color images. Due to the negative direction of the blue body reflection vector, the computed surface reflection components for the blue donut have slightly increased values. However, the surface reflection components for the remaining objects separated out quite well.

Color Figure 27 (see color insert) shows the noise image of the plastic scene. It displays the distance ϵ of every pixel in an image area from the associated dichromatic plane, scaled relative to the width of the planar slices and of the cylinders. Grey pixels represent color pixels that lie on the dichromatic plane. Increased noise values are displayed as increasingly bright or dark pixels, depending on whether the color pixels lie in front of[4] or behind the dichromatic plane. Very bright or dark areas in the

---

[4]The side in the direction of the dichromatic normal.

noise image indicate places where the Dichromatic Reflection Model is not satisfactory in explaining color variation as a mere combination of body and surface reflection with added random camera noise.

Increased noise values occur at pixels on material boundaries where light from two neighboring objects is integrated into the pixel values. The colors of such pixels are a linear combination of the two matte object colors of the neighboring objects. Depending on the orientations of the dichromatic planes, the pixels may be included in either of the two image areas, or they may be left unassigned. Higher noise values also occur at the centers of the highlights due to clipped and bloomed pixels and at internal object edges where the surface orientation changes abruptly. Chromatic aberration causes increased noise values at the highlight on the green donut in the lower right corner. Finally, large noise values exist at places of interreflection between objects, such as on the yellow donut, between the orange and yellow cup and on the right half of the green cup.

5.3   Restoring the Colors of Clipped and Bloomed Pixels

In addition to flagging clipped and bloomed color pixels, the algorithm tries to restore their correct color values. To restore the color values, it exploits the observation that in many cases, clipping and blooming occurs only in one or two color bands, and the pixels may have correct data in the other color bands. The *restoration heuristic* assumes that the smallest of the three values of a color pixel comes from a color band without clipping and blooming. The algorithm uses it to replace the clipped or bloomed pixels by pixels on the matte or highlight line that have the same value in the undistorted band. To decide whether to project a pixel onto the matte line or onto the highlight line, the bloomed pixels are divided into two groups, bloomed matte pixels and bloomed highlight pixels, depending on the reflection line they are closest to. All clipped pixels are projected onto the highlight line.

Color Figures 28 and 29 (see color insert) show the body and surface reflection images of the plastic scene (Color Figure 1) after clipped and bloomed pixels were restored. In the body reflection image, the pixels were restored reasonably well, but there are visible intensity changes between some distorted and undistorted pixels. Such effects can occur if the position of the highlight line is wrong. There is also a limit to the performance that can be achieved: Since all clipped pixels and all bloomed highlight pixels are restored to lie exactly on the highlight line, they all have the same body reflection component. consequently, such pixels cannot exhibit the shading information, which is normally encoded in the distance of a highlight pixel from the highlight line. This may cause a perceptible intensity change between distorted and undistorted pixels in the body reflection image. This data should only be used in combination with the corresponding error image. It may serve as additional, approximate information to deduce the object shape in areas with distorted color data.

The surface reflection image, on the other hand, has gained significantly from the restoring procedure. The clipped and bloomed pixels fill the centers of the highlights with the proper surface reflection color.

For comparison, Color Figures 30 and 31 (see color insert) show the reflection images that result from using the globally determined reflection vectors in Table 5.1. As expected from the discussion on the reflection vectors, the body reflection image displays the blue and yellow donut in black because their reflection vectors are wrong. The surface reflection components of these two areas, on the other hand, are displayed in the object colors blue and yellow. For the remaining objects, the splitting results of the global method are similar in quality to the results of the local method.

5.4   The Use of Intrinsic Reflection Images

The reflection images demonstrate that the color image analysis system described here detects and removes highlights well on a variety of plastic objects. When evaluated qualitatively, the body reflection images generally describe smooth shading across the highlight area. These images may be useful in determining object shape from the shading information [66]. They may also improve the results of algorithms which are currently disturbed by highlights, such as motion and stereo analysis [29, 6]. The surface reflection images show that the highlights from the original image have been detected well. Moreover, the surface reflection images exhibit gradual changes between areas with no surface reflection and areas with very high amounts of surface reflection in the center of the highlights. The images thus provide quantitative information on how much surface reflection exists at every pixel, as opposed to a boolean decision on whether or not a pixel is part of a highlight. The changing amount of the surface reflection component in the highlight on the yellow cup and the small amount of gloss on the handle of the orange cup demonstrate how the surface reflection component changes as a function of the photometric angles. The surface reflection images may be useful input for methods that determine object shape from highlights [58].

To illustrate the changing amounts of body and surface reflection more quantitatively, Color Figure 32 (see color insert) displays the reflection pro-

files for a row of pixels in the body and surface reflection images (Color
Figures 28 and 29). It also shows the amounts of noise, ϵ, at these pixels.
The selected row runs across the yellow donut, the orange cup and the
yellow cup. The body reflection profile shows that the highlights have been
removed reasonably well from the objects. Some error was introduced in
the body reflection components of clipped and bloomed pixels whose colors
were restored. The surface reflection profile consists of a series of spikes at
the positions of the highlights on the objects. Each spike has the form of a
peak which falls off to the sides, demonstrating the quantitative nature of
the highlight analysis of this approach. A few small peaks exist at object
boundaries. These are pixels on material boundaries where light from more
than one object is mixed into a single pixel value. Since such pixels generally do not lie close to the matte line, they generate a disproportionately
high surface reflection component.

5.5   Summary

This chapter has shown how the Dichromatic Reflection Model can be used
to split color pixels into their body and surface reflection components. Two
methods for determining the body and surface color vectors of a segmented
image area were described and compared. The first method uses a global
analysis of the histogram generated by projecting all pixels from an image
area into the color cube. The second method exploits the information on
local color variation that is already available from the image segmentation described in the previous chapter. The global shape analysis method
performs well when the color clusters have long, well-populated linear subclusters. The local method, on the other hand, depends very little on the
length of the subclusters or on spurious noise caused by few pixels. It
can be biased, however, by structural irregularities, as caused by a significant amount of pixels that are often related to the systematic influence of
physical processes beyond the scope of the current model.

The basic algorithm of splitting color pixel values of an image area
into their reflection components consists of a linear transformation from
the original color coordinate system, (R, G, B), into a coordinate system,
$(\overline{C}_b, \overline{C}_s, \overline{C}_b \times \overline{C}_s)$, defined by the body and surface reflection vectors.
The method is augmented by heuristics to restore the colors of clipped and
bloomed pixels which do not follow the dichromatic theory. The results
show that the algorithm separates the reflection components very well.
The generated reflection images may be useful for other areas of computer
vision which cannot handle the simultaneous occurrence of shading and
highlights.

6

Results and Discussion

The previous chapters have presented a color image understanding system that exploits the Dichromatic Reflection Model. The description of the system was accompanied by a few exemplary results that demonstrated how the algorithm performed on real images. This chapter presents more results and discusses the strengths and limitations of this approach.

6.1 Further Results

The algorithm has been tested on a series of color images of scenes with dielectric objects under varying illumination conditions. The upper left quarters of Color Figures 33, 34, and 35 (see color insert) show the images of three plastic cups under yellow, white and pink light.

The upper right quarters of the figures display the segmentations that were generated by the algorithm presented in Chapter 4. The images demonstrate that the method performs well on a variety of dielectric objects under varying illumination conditions. In all images, the segmentations outline the material boundaries of the objects very well. Because the algorithm models matte and highlight color variations as optical reflection processes, it ignores color changes along highlight boundaries, and it also tolerates shading changes on the objects.

Table 6.1. Normalized body and surface reflection vectors of the three plastic cups under yellow light (local method).

	body reflection vector	surface reflection vector
yellow cup	(0.85,0.52,0.08)	(0.79,0.59,0.17)
orange cup	(0.97,0.23,0.08)	(0.74,0.65,0.18)
green cup	(0.39,0.91,0.15)	(0.70,0.68,0.20)
illumination vector:		
- average surf. refl.		(0.74,0.64,0.18)
- independent meas.		(0.73,0.66,0.19)

Tables 6.1 through 6.4 present the body and surface reflection vectors of the images in Color Figures 33 through 35 that were determined by analyzing local color variation in the image. Tables 6.5 through 6.7 show the results of using the global method for estimating the reflection vectors from the entire cluster shape. Qualitatively, the body reflection vectors in all tables show the correct trends: The body reflection vectors of the green cup are dominated by the green component, while the body reflection vectors of the yellow and orange cups are dominated by the red component. Under yellow illumination, the body reflection vectors have smaller blue components than under white light. Under pink light, on the other hand, the green components of the vectors are smaller than under white light. Currently, no independent measurements of the body reflection colors (e.g., by a spectroradiometer) are available to enable a quantitative evaluation of the results.

In every table, the surface reflection vectors of the three cups are approximately constant. This supports the observation that the three highlight clusters are parallel to one another. This fact is especially visible in Tables 6.1, 6.2 and 6.5, which show the results of the local analysis of the cups

Table 6.2. Normalized body and surface reflection vectors of the three plastic cups under white light (local method).

	body reflection vector	surface reflection vector
yellow cup	(0.80,0.58,0.15)	(0.50,0.53,0.68)
orange cup	(0.95,0.27,0.16)	(0.47,0.58,0.66)
green cup	(0.20,0.92,0.33)	(0.43,0.54,0.73)
illumination vector:		
- average surf. refl.		(0.47,0.55,0.69)
- independent meas.		(0.58,0.57,0.58)

Table 6.3. Normalized body and surface reflection vectors of the three plastic cups under pink light (local method).

	body reflection vector	surface reflection vector
yellow cup	(0.92,0.38,0.06)	(0.84,0.49,0.23)
orange cup	(0.98,0.19,0.06)	(0.79,0.53,0.30)
green cup	(0.72,0.63,0.29)	(0.89,0.42,0.15)
illumination vector:		
- average surf. refl.		(0.84,0.48,0.23)
- independent meas.		(0.71,0.59,0.38)

Table 6.4. Normalized body and surface reflection vectors of the three plastic cups under pink light (local method).

Table 6.5. Normalized body and surface reflection vectors of the three plastic cups under yellow light (global method).

	body reflection vector	surface reflection vector
yellow cup	(0.85,0.52,0.08)	(0.71,0.68,0.20)
orange cup	(0.96,0.27,0.09)	(0.67,0.72,0.17)
green cup	(0.39,0.91,0.16)	(0.72,0.66,0.20)
illumination vector:		
- average surf. refl.		(0.70,0.69,0.19)
- independent meas.		(0.73,0.66,0.19)

Table 6.6. Normalized body and surface reflection vectors of the three plastic cups under white light (global method).

	body reflection vector	surface reflection vector
yellow cup	(0.80,0.58,0.16)	(0.25,0.46,0.85)
orange cup	(0.95,0.26,0.15)	(0.43,0.59,0.68)
green cup	(0.21,0.92,0.34)	(0.44,0.47,0.77)
illumination vector:		
- average surf. refl.		(0.38,0.51,0.77)
- independent meas.		(0.58,0.57,0.58)

Table 6.7. Normalized body and surface reflection vectors of the three plastic cups under pink light (global method).

	body reflection vector	surface reflection vector
yellow cup	(0.93,0.37,0.05)	(0.71,0.62,0.34)
orange cup	(0.98,0.20,0.05)	(0.77,0.55,0.32)
green cup	(0.69,0.67,0.28)	(0.87,0.45,0.20)
illumination vector:		
- average surf. refl.		(0.79,0.54,0.29)
- independent meas.		(0.71,0.59,0.38)

under yellow and white light and of the global analysis of the cups under yellow light. Normalized estimates for the yellow, white and pink illumination were obtained by taking an image of a grey chart under such lights, as described in Section 5.1.3. They are shown at the bottom of the tables, along with the average surface reflection vector which the system uses as its internal estimate of the illumination color. The independently measured illumination vectors are generally close to the fitted surface reflection vectors and to the averaged vector. This demonstrates that the surface reflection components of dielectric materials can be determined from real images to provide an estimate of the illumination color – information needed in color constancy algorithms [33, 109, 119].

The lower left quarters of Color Figures 33 through 35 show the intrinsic body reflection images of the three cups under yellow, white and pink light. The lower right quarters show the corresponding surface reflection images. The images were generated using the local estimates of the reflection vectors that are shown in Tables 6.1 through 6.4. For comparison, the corresponding results using the global method (on hand-segmented images) can be found in [99]. The intrinsic images demonstrate that the image analysis system described here detects and removes highlights well on a variety of dielectric materials under varying illumination colors. When evaluated qualitatively, the body reflection images generally describe smooth shading across the highlight area. The surface reflection images show that the highlights from the original images have been detected well. The reflection images may be a useful input to methods that determine object shape from shading [66] or highlights [58].

6.2 Comparison with a Traditional Color Segmentation Algorithm

Color Figure 2 shows the results of applying a traditional color segmentation method to Color Figure 1. As described in Chapter 1, it was obtained by running Phoenix [165, 146] on three user-selected features: intensity, hue and saturation. Some of the highlights are separated from the surrounding matte object parts while other highlights are integrated with the matte areas. In addition, the matte areas are sometimes split into dark and bright areas. This can be seen on the right half of the green cup, as well as on the dark red donut and on the green donut.

It is not easy to predict how Phoenix will perform on a given image, because its decisions are not directly related to the influences of physical processes. Phoenix uses a pre-defined, fixed set of color features. It measures color variation only in relation to them and segments image areas using only one feature at a time. Projecting the color information from a three-dimensional color space to one-dimensional histograms, Phoenix looses most of the information about the three-dimensional structure of the color variation. Furthermore, the histograms are related to a fixed set of pre-defined color properties, and they are generally defined in relation to the origin in the color space, i.e., the black corner of the color cube. Under ambient light, the color clusters are translated such that they converge toward a brighter point in the color space [25, 163, 187] – an effect that is not anticipated in the predefined list of color properties. The properties are also generally defined in relation to white illumination. Under a different (e.g., yellow) illumination color, features like intensity, hue and saturation do not define a conical coordinate system with the illumination vector at its center, as might be intuitively expected. For these reasons, the feature axes do not necessarily reflect the structure of the color cluster.

Compared to Phoenix, the color cylinders and dichromatic planes of the approach described here capture color variations in a much more flexible way. The color cylinders and planes of the linear and planar hypotheses are determined by analyzing color clusters in their three-dimensional structure. Their position and orientation is independent of a chosen coordinate system in the color space, as well as of a user-defined set of features. Consequently, linear and planar hypotheses can be expected to capture color variation on objects much better and to generate better segmentations of the objects, possibly even under ambient light. This is evident when Color Figure 2 is compared to Color Figures 18, 19, and 20. Since the linear, planar and final segmentations account for shading variation on matte object areas, for highlight reflection, and finally for camera limitations, they include all

matte and highlight pixels, as well as bloomed and clipped pixels, into a
single object area.

6.3   Control Parameters

So far, this chapter has presented the promising results of the analysis of
a series of color images under several illumination colors under the guidance of a physical reflection model. The following sections will discuss the
system's dependence on control parameters and heuristics, as well as on
limiting assumptions in the dichromatic theory. The system uses a few
control parameters and simplifying heuristics in its color image analysis.
Most of them have been mentioned at the appropriate places in the previous chapters. They are related to camera limitations, to the limitations of
color information (e.g., in dark image areas) and to the basic assumptions
of the Dichromatic Reflection Model. This section shows the influence of
the parameter settings on the system's performance in analyzing the color
images shown above. The following section will discuss the influence of the
heuristics.

The control parameters, as well as the parameter values that were chosen
to generate the above mentioned results, are listed in Table 6.8. It seems
that most of the parameters are related in quite an intuitive way to the
camera limitations and to the models of the reflection processes and are
thus relatively easy to set (better than in many other algorithms where
control parameters are not related in an intuitive way to scene properties).

6.3.1   Clipping Threshold, Noise Density and Camera Noise
The clipping threshold and the blooming noise density parameter are
related to the the limited dynamic range of cameras. The density parameter expresses the blooming heuristic, classifying color pixels as noisy or
bloomed pixels, if they occur only once in an object area (i.e, if only one
pixel falls into that color bin in the color space).

The variance σ_0 describes the estimated noise of the camera. The camera
noise was estimated by repeatedly taking a black-and-white picture of a
white board, measuring the intensity of the center pixel and then computing
its statistics. σ_0 corresponds to twice the value of the standard deviation σ
of these measurements such that the integral of the Gaussian curve in the
range $[\mu - \sigma_0, \mu + \sigma_0]$ describes about 96% of all measurements.

Since σ_0, the clipping threshold and the noise density parameter are
related to the characteristics of the camera, and since all images shown

Table 6.8. Control parameters of the algorithm.

	plastic scene	3 cups yellow light	3 cups white light	3 cups pink light
clipping threshold	240	240	240	240
blooming noise density	1	1	1	1
camera noise σ_0	2.5	2.5	2.5	2.5
smoothing window	51	51	51	51
fitting tolerance for recursive line splitting	21	21	21	21
minimum intensity	23	30	30	30
cylinder radius, width of planar slice	$4\sigma_0$	$4\sigma_0$	$6\sigma_0$	$4\sigma_0$
initial window size	10 x 10	10 x 10	10 x 10	15 x 15
minimum area size	500	500	500	700

here were taken with the same camera, these parameters are constant in
the analysis of the given images.

6.3.2   Smoothing Window and Fitting Tolerance for Global Shape Analysis

The recursive polygon-fitting procedure of the global method to estimate
the reflection vectors (Section 5.1.1) requires two parameters: The size of
a smoothing window which determines how many neighboring entries in
the distance function $d(s)$ influence the averaging operation and a distance
threshold which controls the recursive splitting operation. Neither of these
two values has ever been critical to any results.

6.3.3   Minimum Intensity

The remaining four parameters are more critical to the performance of
the algorithm. Since all matte clusters converge and overlap at the dark
corner of the color cube, color information cannot be used reliably in dark
image areas. To avoid segmentation conflicts, dark image areas are excluded by setting a threshold for minimum intensity. However, Section 4.4
pointed out that the amount of overlap between the color clusters of neighboring objects depends on the angle between the respective body reflection
vectors. Thus, a chosen threshold may not resolve all color conflicts and,
as a result, an image area may bleed into a neighboring one. The amount
of bleeding can be reduced by increasing the required minimal intensity. A

very high threshold, however, may cut off so much of the color cluster of a dark object that the body reflection vector can no longer be estimated reliably.

Color Figure 36 (see color insert) demonstrates the influence of the minimum intensity parameter on the example of the three plastic cups under yellow light: If a low threshold (20) is selected, the yellow cup bleeds into the neighboring orange one. When the orange hypothesis is analysed afterward, it regains most of the pixels that were wrongly assigned to the yellow cup, due to the proximity heuristic of assigning color pixels to the closest color vector (see Section 4.4). Many very dark pixels at the boundary of the orange cup, however, now have an arbitrary assignment to either the yellow or the orange cup. This is shown in the upper two quarters of Color Figure 36. If, on the other hand, a high threshold (50) is chosen, the analyzable area on the green cup becomes quite small, as shown in the lower left quarter of Color Figure 36.

This parameter touches the limitations of the analysis of color variation in images. To overcome the problems of bleeding, a geometric analysis that relates object shading to object shapes is necessary. Assuming smooth object shapes, the shape information can be used to extrapolate the image analysis into areas that were excluded by the dark heuristic or in which bleeding occurred.

6.3.4   Cylinder Width

Another important parameter in the algorithm is the width of the matte cylinders and the planar slices. Due to secondary effects, such as interreflection, on color variation in real images, the linear color clusters do not always have a small, circular cross-section. Furthermore, the clusters may not even be really linear, due to pigment variations in the material (e.g., dirt on the object) or to minimal amounts of interreflection and other illumination changes. For such reasons, and also because of possible estimation errors in the direction of the color vector, the system needs to use fairly thick cylinders and planar slices to include most of the pixels on an object into the segmented area. The drawback of unnecessarily thick cylinders and planar slices is a loss of sensitivity to the color properties of the image. For example, a thick cylinder width generates more bleeding into neighboring objects, as discussed in Section 6.3.3.

Color Figure 37 (see color insert) shows that the yellow cup under white light requires a thick cylinder of $6\sigma_0$ instead of the $4\sigma_0$ used for the other images. The upper left quarter shows that the matte color cluster of the cup is not linear. This may have been caused by slight amounts of interreflection from the orange cup onto the dark pixels in the lower half of the yellow

cup. If a matte cylinder width of $4\sigma_0$ is chosen, this interreflection biases
the orientation of the yellow body reflection vector such that the matte
cylinder fits well for the dark yellow pixels, as shown in the lower left quarter
of Color Figure 37. Bright matte pixels around the highlight, however,
are not contained in this cylinder. Such pixels then generate another,
slightly different, linear hypothesis, as shown in the lower right quarter.
Accordingly, the segmentation for the yellow cup (in the upper right quarter
of the figure) is divided into two major areas, one containing the brighter
upper part of the cup (shown in pink), and the other containing the darker
lower part (shown in red) where interreflection from the orange cup changes
the yellow object color.

The system currently does not have the capabilities to combine several
very similar hypotheses of neighboring object areas. It lacks an error recovery mechanism that reconsiders previous hypotheses. Reconsideration
capabilities require a mechanism to detect and resolve conflicts between sets
of hypotheses. Depending on the circumstances, such competing hypotheses may need to be combined into a single hypothesis, as shown above, or
they may lead to refinements for some of the hypotheses, possibly providing
an increased sensitivity to distinguishing features.

It may also be necessary to combine the analysis of local color variations
with an analysis of the global cluster shape. Such a global analysis has
to include components for determining the optimal cylinder size and the
orientation for a linear color cluster, as well as methods for determining
irregularities in the cylindrical cluster shape due to interreflections and
other secondary effects. Some work on globally analyzing the color variation in an entire image exists or is in progress [159, 193]. These techniques
use a five-dimensional vector space, representing three color parameters,
R, G, and B, and two positional parameters, x, and y. They use pattern
recognition techniques to determine clusters in this feature space. It may
be interesting to study whether and how such techniques can be used in
the local approach to image analysis of this text to determine the linear
subclusters of a color cluster from one object.

6.3.5   Initial Window Size

The segmentation algorithm depends on the window size in the initial-
ization step. If the windows are too large, many of them overlap several
objects, and many local color clusters are classified as volumes. If, on the
other hand, the windows are too small, color variation on a relatively flat
or dark object may not be detected and the color clusters of windows on
such objects will then be classified as pointlike clusters.

This is shown in the initial classification of the cups under pink light in Color Figure 38 (see color insert) for 10 × 10 windows. Since green and pink are approximately complementary colors, the green cup appears quite dark under pink light. Consequently, the local shading changes on the cup are very small. They are not large enough in 10 × 10 windows for the color clusters to be classified as linear clusters.

When the window size is either too small or too large, the intitialization step does not provide sufficient information for the subsequent phases of the algorithm to select and apply hypotheses about linear and planar color variations. The system may be able to determine an optimal window size by generating descriptions of local color variations at several window sizes and then selecting the window size that resulted in the highest percentage of linear and planar windows.

6.3.6 Minimum Area Size

One very crucial heuristic in the system described here assumes that the selected linear hypotheses all represent matte clusters. In principle, linear color clusters can also exist in highlight areas or at material boundaries. The heuristic is based on the observation that the largest linear clusters are generally matte clusters, because highlights are generally much smaller than matte object areas. Similarly, linear clusters at material boundaries occur very rarely and are generally confined to a very small region around the boundary. To select only matte regions as linear hypotheses, the algorithm inspects color clusters by decreasing size of the corresponding regions in the image, and it only selects regions above a threshold. However, if an object is quite flat, and if the highlight on such an object is not clipped and bloomed, it can result in a linear color cluster of a significant size and mislead the system, which will classify it as a matte area and then look for a neighboring highlight.

This occurs in the image of the three cups under pink light, for a minimum area size of 500. Color Figure 39 (see color insert) shows that the pink highlight on the green cup subtends a large area and is selected as an independent matte hypothesis.

6.4 Simplifying Heuristics

The above results show that the approach to color image understanding presented here can be used to segment and analyze a series of color images very successfully. In applying the dichromatic theory to color images, however, the system relies on a few simplifying heuristics. They are related to the problem of determining the structure of a color cluster. The system currently works with a very rigid, built-in model of what color cluster shapes to expect, assuming that clusters always consist of two linear subclusters that form a skewed T. This section discusses these heuristics.

6.4.1 Matte Heuristic: Large Linear Clusters Are Matte Clusters

As discussed in Section 6.3.6, the segmentation algorithm assumes that linear hypotheses from large image areas describe matte pixels on an object. This heuristic depends on the curvature on the objects in the scene, as well as on their distance from the camera. Future work may involve analyzing linear clusters both as potential matte clusters and as potential highlight clusters.

6.4.2 Highlight Heuristic: Color Clusters Have Only One Highlight Branch

When the system selects a surface reflection vector from a list of highlight candidates in Section 4.5, it chooses the cluster that is closest to the average starting point of the candidates for highlight clusters. This step implicitly assumes that the clusters of all highlight candidates are part of the same highlight cluster. This may not be the case for concave objects. The two planar faces in Figure 6.1 each reflect a highlight into the camera. Due to the different illumination conditions, the incidence angle between the illumination ray and the surface normal is larger at point P_1 than at point P_2. Accordingly, the body reflection component under the highlight at P_1 is smaller than the body reflection component under the highlight at P_2, and the two highlights generate separate highlight clusters in the dichromatic plane.

The limiting case is produced by a concave ellipsoid with the light source and the camera positioned in the focal points of the revolving ellipse, as shown in Figure 6.2(a). Since every point on the surface of an ellipsoid has a surface orientation that obeys the laws of perfect mirror reflection between the two focal points, the entire image of the ellipsoid consists of one big highlight. The angle between the local surface normal and the incident ray varies. The amount of body reflection varies accordingly over the ellipsoid. Since every point on the ellipsoid exhibits maximal surface reflection, the resulting color cluster is a straight line in the direction of the color of body reflection, but translated to higher intensities according to the amount and color of surface reflection.

If the surface of the ellipsoid consists of small planar patches, as in Figure 6.2(b), the body reflection is approximately constant on each patch,

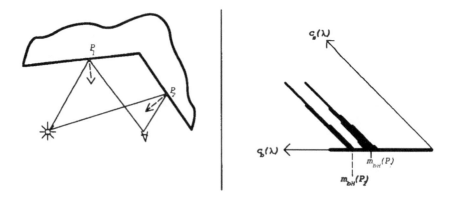

Figure 6.1. Cluster shapes for concave objects.

whereas the surface reflection varies. The color cluster then reveals approximately the shape of a parallelogram with the sides in the direction of the body reflection and surface reflection vectors.

This analysis shows that a color cluster may have several highlight clusters. Future work will be necessary to improve the criteria of determining the starting point of a highlight line.

6.4.3 50%-Heuristic

The 50%-heuristic also makes limiting assumptions on the scene. So far, it has only been evaluated for spherical objects that were illuminated and viewed from the same distance. Different conditions may exist on cylindrical, ellipsoidal or planar objects. Furthermore, the heuristic does not apply to concave objects since, on such objects, the brightest visible matte point may lie on one surface while the highlight is on another surface with different illumination conditions. The highlight may thus start at an arbitrary place on the matte line.

6.5 Limitations of the Dichromatic Theory

The heuristics and parameters discussed above limit the performance of the algorithm in a practical way. They are related to implementational decisions. This section discusses principle limitations of the dichromatic theory.

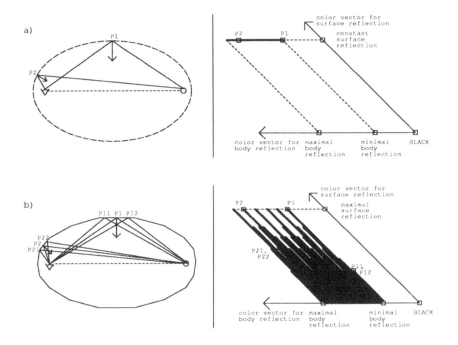

Figure 6.2. Light reflection from a concave ellipsoid.

6.5.1 Collinear Matte Clusters

This approach is unable to find material boundaries between neighboring objects with collinear matte clusters because all such color variation falls within a single cylinder. The system will need to be extended by a geometric analysis, linking intensity gradients to object shape, to distinguish between such objects. A geometric analysis will also be needed to analyze dark image areas which are currently excluded because their color information is too unreliable.

6.5.2 Color Difference between Body and Surface Reflection

The current method to split color pixels into their reflection components (but not the segmentation) relies on a characteristic color change between the matte object color and the highlight color. There needs to be a minimum angle between the orientations of the body and surface reflection vectors of an object. How big the angle needs to be depends on the estimated camera noise (as related to the width of the matte cylinder). If the matte and highlight clusters are approximately collinear, the reflection components cannot be separated well. This is the case if an object has the same color as the light source or if the object is a uniform reflector, such as grey or pale pastel objects. It also occurs for a narrow-band illuminant. In the extreme, monochromatic light is reflected in the same monochromatic color (in varying amounts) from every object in the scene – independent of the object color.

Color Figure 33 shows an example of such limitations. The body reflection image of the yellow cup under yellow light is quite coarse. The salt-and-pepper noise on this cup is due to the approximate collinearity of its body and surface reflection vectors. Small amounts of noise in its color pixels have a large effect on the computed amounts of underlying body reflection.

6.5.3 Missing Linear Subclusters: Special Materials

The algorithm also has problems when one of the two linear clusters does not exist or is very small for a specific object.

The matte cluster is missing if the viewed object is very dark or if the scene is illuminated with a narrow-band illuminant that does not overlap with the wavelengths at which the material reflects light. Matte clusters also do not exist for metallic objects.

On the other hand, the highlight cluster may be missing if an object does not reflect a highlight into the camera due to its position in the scene and the illumination geometry.

A third case arises from objects with very rough surfaces such that every pixel in the image area has both a significant body and surface reflection component. The color cluster may then fill out the entire dichromatic plane. A common special case of this phenomenon are so-called "matte materials" – as opposed to glossy materials – which reflect a constant amount of surface reflection in every direction and thus never exhibit a highlight in the common sense of the word. The corresponding color clusters are linear clusters, translated from the origin of the color space according to the constant surface reflection component.

The system is currently not capable of distinguishing between all these cases. Problems may arise when a scene contains both metallic and dielectric objects: The algorithm then needs to determine whether a single linear cluster in the color cube should be interpreted as a highlight cluster generated by a metal, or as a matte cluster generated by a dielectric object without a highlight, or as a mixture of both reflection processes on "matte" materials. In addition to exploiting previously determined scene properties, such as the illumination color, the system will need to analyze the intensity gradients along the axes of the linear color clusters and relate them to the properties of m_s and m_b, as described in a geometric model of light reflection. One viable approach might be to use shape from shading to generate the shape of a hypothetical dielectric object and to compare it with the shape of a hypothetical metal, as generated by a shape from highlights method.

6.6   Summary

When compared to the segmentations produced by traditional color image segmentation approaches, the results presented in this chapter indicate that the color image segmentation described in this text is able to segment color images very well ignoring color variation due to shading. In addition, the system splits the images into their reflection images. The results may be a useful starting point for further image analysis, such as determining object shapes and the illumination configuration from the intrinsic images and from the hypotheses. Several control parameters and simplifying heuristics, as well as limiting assumptions in the reflection model, currently limit the applicability of the method, yet there are ways to relax such limitations. This chapter has indicated some methods to achieve such improvements.

7

Summary and Conclusions

This text describes a physical approach to color image understanding. It uses an intrinsic reflection model that accounts for the effects of shading and highlight reflection in color images. Combining this model with a camera model and with a geometric analysis of the relationship between object shapes and color cluster shapes, this work is able to demonstrate that a theoretical, physical model can be transformed into a very successful computational procedure for analyzing color images. The resulting color image interpretation system segments color images along material edges - ignoring color changes due to highlights and shading - and separates the images into their intrinsic reflection components.

7.1 Contributions

This research impacts on several aspects of computer vision, especially color image understanding.

- **We use a physical model of shading and highlight reflection for color image segmentation and analysis.**
  By using this intrinsic reflection model, the work stands in contrast to traditional, statistical color image segmentation methods which

assume that object color is a constant property of an object and which therefore attribute color variation to sensor noise and material changes.

- **The algorithm is better than statistical segmentation.**
  Because this work models physical effects of the scene, it is able to produce image segmentations that ignore color changes due to highlights and shading, while traditional segmentation systems generally draw boundaries between matte and highlight areas on objects and sometimes even divide matte object areas into darker and brighter areas. Furthermore, the system described here can adapt its analysis to varying illumination conditions, such as different illumination colors and ambient light. In comparison, segmentation criteria in traditional algorithms are based on rigid, user-defined color features which generally assume white illumination and no ambient light.

- **It demonstrates the applicability of a physical reflection model to color image understanding.**
  Intrinsic models have been developed for quite a while, and, although their relationship to physical properties of the scene was intriguing, there were doubts as to how effectively they could be used in algorithms. This text demonstrates that it is possible to analyze real color images by using a physics-based color reflection model.

- **It shows the importance of performing the low-level image analysis process in direct interaction with the image data.**
  Traditionally, image analysis has been designed as a hierarchy of steps from which only those on the lowest level access the original image data, while the steps on higher levels analyze the results provided by the lower levels. This traditional approach benefits from information reduction, but it has to cope with problems arising from insufficient performance of the methods at the lower levels, such as missing or erroneous edges in an edge image. This is a difficult problem, since the lower-level image analysis methods are generally based on a heuristic or statistical analysis of image properties instead of on a model describing physical properties of the scene. In contrast, the approach described here accesses the original image data at all stages of its analysis. It frequently alternates between bottom-up and top-down analysis to generate hypotheses about physical processes and reapply them to the image. This generate-and-test

approach may be extensible to a more complete low-level image understanding expert system for physical segmentation.

- **This work demonstrates the value of separately modeling multiple, simultaneous physical effects.**
  Previous work with intrinsic models concentrated on interpreting the influence of only one optical effect at a time or combined the influences of several effects into a single representation, such as a reflectance map. The assumption of orthographic projection and illumination of reflectance maps significantly limits their applicability to scenes with highlights: In laboratory scenes the camera and light source are relatively close to the objects, so the strength of surface reflection at a surface patch on an object does not depend only on the orientation of surface normal of the surface patch but also on its position in the scene. By modeling shading and highlights as distinct effects, this work is able to separate their influences into two intrinsic images, as well as to distinguish them from material changes. The intrinsic images may then be analyzed separately, as two independent sources of information constraining each other. This approach does not assume orthography for modeling any optical effect. It may open the way for analyzing highlight reflection under perspective projection on the intrinsic surface reflection image, while using traditional Lambertian reflectance maps to analyze shading in the body reflection image.

- **The intrinsic reflection model provides a rational, physical basis for understanding the capabilities and limitations of the approach.**
  Most traditional approaches do not relate their assumptions directly to scene properties and are difficult to analyze, extend, and control.

  - **Since the control parameters of the new algorithm are related to physical scene properties, there is an objective criterion for setting these values.**

  - **The algorithm produces useful results.**
    In addition to providing an image segmentation, the algorithm described here generates intrinsic reflection images and a set of hypotheses that describe object colors and the color of highlights and the illumination. The intrinsic reflection images have a simpler relationship to the illumination geometry than the original image and may thus improve the results of many other computer vision algorithms, such as motion analysis, stereo vision, and shape from shading or highlights. The highlight image

may also provide strong evidence for the position of the light
source. The hypothesized illumination color is useful for color
constancy algorithms. Alltogether, the information provided by
the algorithm may be a useful starting point for further analysis
to determine object shapes through a geometric image analysis
and for material classification. The algorithm can thus be seen
as the first step in a physics-based image analysis system.

- The complexity involved in extensions necessary to include further physical processes can be estimated.

- The work demonstrates that highlights are a combination of
  two optical processes, body and surface reflection.

    - It points out that highlights have a body reflection com-
      ponent.
      With very few exceptions, highlights on objects have so far been
      viewed as areas on objects in which the object color is replaced
      by (not mixed with) a highlight color. The color clusters and the
      body reflection images in this work show that the shading pro-
      cess also exists under the highlight and that it can be separated
      from the superimposed surface reflection.

    - It models highlight variation quantitatively.
      The Dichromatic Reflection Model and the color histograms
      demonstrate that, in real scenes, a transition area exists on the
      objects from purely matte areas to the spot that is generally
      considered to be the highlight. This transition area determines
      the characteristic shapes of the color clusters which is the infor-
      mation that the system uses to distinguish highlight boundaries
      from material boundaries and to detect and remove highlights.
      This view of highlights should open the way for quantitative
      shape-from-gloss analysis, as opposed to binary methods based
      on thresholding intensity.

- The text develops a radiometric camera model and demonstrates its influence on color image variation.
  By analyzing real color images that were taken in a controlled environment[1], this work has been able to relate failures of early versions
  of the algorithm to a few, specific properties of the scene or camera.
  Such problems could then be studied systematically and their impact
  on real color images could be demonstrated. In most cases, effects

---

[1] The Calibrated Imaging Laboratory at Carnegie Mellon University.

related to camera problems could then be accounted for in a camera model. Such an analysis of unexpected effects is much more difficult to pursue on images taken in an unconstrained scene, such as outdoor images. The experiences and methods gained can be of practical use to the entire vision community.

7.2 Directions of Future Research

The work presented in this monograph is a promising starting point for a new approach toward color image understanding. It opens the way for several directions of future work.

7.2.1 Extensions of the Model

This work demonstrates that the Dichromatic Reflection Model contributes to understanding color images. However, the model makes several simplifying assumptions which restrict the illumination conditions and the materials in the scene. Because of these assumptions the model can currently only be used to understand scenes taken under very restricted conditions. The results of using the reflection model in analyzing real images, such as the image in Color Figure 1, have shown that even images of carefully selected scenes often contain areas in which the model does not apply (see Color Figure 27). To provide a broader applicability of the method, the model needs to be extended to account for more physical processes, especially for more general illumination conditions. It will thus be necessary to study and model such physical effects, to describe and verify their behavior in images, and to design algorithms that account for them.

Limiting assumptions on illumination conditions have proven to impose the strongest restrictions on the performance of the algorithm. Interreflection and shadow casting between objects exist in nearly every moderately complicated scene, such as the image in Color Figure 1. Furthermore, many realistic scenes will contain more than a single, point-like light source. There will generally be some amount of ambient light and possibly even several light sources that shed differently colored light onto the scene. A more generally applicable algorithm will need to account for changing illumination properties in the scene. Future work will involve investigating the properties of such illumination conditions. It seems appropriate to relax the illumination constraints gradually, starting with the analysis of ambient light. Ambient light has been described by a constant additional term in some reflection models [163, 25]. Tong and Funt have presented promising first results on analyzing real images with ambient light [187], and Section 6.2 of this text indicates the potential of the system described here for handling such images. It now needs to be verified in real images.

The Dichromatic Reflection Model also restricts the material types that can be analyzed. It assumes that all objects are made of dielectric non-uniform opaque materials with evenly distributed pigments in the material body. This assumption applies to many man-made products, such as kitchen tools, toys, office supplies and painted cars. However, it excludes metals. It also excludes objects with color texture patterns that are very small or change gradually in such a way that they cannot be described as an assembly of separate materials along the object surface. Many industrial vision applications need an analysis of metallic objects. This thesis contains some discussion on the properties of metallic color clusters in real images and how they can be distinguished from dielectric color clusters. Further suggestions have been made by Healey, Binford, and Blanz [59, 62]. More rigorous investigations of this subject will be necessary in the future.

The 50%-heuristic describes a first step toward quantitatively relating object shapes and the illumination set-up to the color cluster shapes. However, it discusses only the reflection behavior on a sphere with the illumination source and the camera positioned at the same distance from the object center. Further investigations will be necessary to discuss the color cluster shapes from more general illumination geometries and from other object shapes, such as cylindrical, ellipsoidal or planar objects and from concave objects.

7.2.2   Applications

The color analysis program generates several kinds of results which can be further analyzed. The two intrinsic reflection images provide separate descriptions of the shading component and the highlight component at every pixel in the image. In the discussion of the Dichromatic Reflection Model, the body reflection component was related to matte shading on objects. At a qualitative level, the intrinsic body reflection images seem to support this expectation. However, this claim needs to be verified. It depends on the properties of the material body, which the model assumes to be describable by a random distribution of the pigments.

If the body reflection image does describe object shading, a shape-from-shading algorithm can be used to extract object shape from the body reflection image. Similar results can be obtained in the highlight areas by applying a shape-from-highlights algorithm to the surface reflection image. This shows that two independent sources for generating object shape information exist, the results of which need to be compared and combined into one consistent, shape interpretation.

Assuming smooth object shapes, the shape information can be used to extrapolate the image analysis into areas that were previously excluded or badly analyzed, because the pixel colors were too dark or clipped or noisy. Such a geometric analysis of the scene can be used to separate neighboring objects with collinear matte color clusters, as well as to find edges on an object that exhibit shading boundaries because the surface orientation changes abruptly. The analysis may also be able to consider trouble spots caused by color bleeding, shadow casting and interreflection.

The geometric analysis may also be able to perform material classifications for objects with a single linear color cluster which can be the highlight cluster of a metal or the matte cluster of a dielectric. Pixels of such a cluster do not possess varying amounts of both reflection components. The color variations either correspond entirely to changing amounts of body reflection or to changing amounts of surface reflection. The correct nature of the color variation can be determined by testing both interpretations in parallel, analyzing the color variation both by a shape-from-shading method and by a shape-from-highlights method and choosing the more likely shape interpretation.

The highlights also provide useful information about the illumination. The color of the highlights on dielectrics can be used to determine the color of the light source. Such information is useful in color constancy algorithms to discount the influence of the illumination on the perceived object color [33, 109, 119]. Future work may investigate how well the estimated illumination vectors of the color analysis program can be used for this purpose.

The positions of highlights on the objects also constrain the illumination geometry. Some such information is, in principle, available from the shape of the color clusters, i.e, from the starting point of the highlight cluster. The positioning of the start of the highlight line in the color analysis algorithm is very susceptible to noise, however, and may not be a reliable source of information. Furthermore, there is not a one-to-one mapping between the start of the highlight line and the photometric angles. Future work may involve investigating whether there are ways to generate hypotheses about the shapes of objects from shading and highlight information while simultaneously estimating the illumination geometry

Finally, the intrinsic noise and error images generate hypotheses about the places and extents of image areas that encounter camera limitations. Future work may try to use these images to adjust the camera aperture automatically such that the dynamic range of the camera is optimally used.

Although the current method has only been applied in a laboratory setting, its success shows the value of modeling the physical nature of the visual environment. This text presents an important step towards that

8

Related Work from 1988 until 1992

The work presented in the previous chapters was concluded in 1988 [93, 99, 100]. As one of the pioneering, physics-based computer vision systems to exploit optical properties of materials, it touches on several topics that have since evolved into exciting research areas. Physics-based research had also been pursued before 1988. Especially the seminal early work on shape from shading and photometric stereo by Horn, Ikeuchi and Woodham [66, 69, 74, 76, 209] has provided the solid foundations to analyze the challenging problems in physics-based vision in a coherent and focussed manner.

Recently, activities in physics-based vision have rapidly increased in number, and the field has gained significant visibility. For example, the Marr Prizes of the last two International Conferences on Computer Vision (ICCV 1988 and 1990) as well as one of the two Outstanding Paper Awards at the European Conference for Computer Vision (ECCV 1992) were awarded for research in physics-based computer vision [44, 113, 134]. An entire session at the DARPA Image Understanding Workshop in 1989 was devoted to this research area; a special issue of the IEEE Transactions on Pattern Analysis and Machine Intelligence (PAMI) covered the topic in 1991 [89]; and a symposium on physics-based computer vision was held in conjunction with the IEEE Conference on Computer Vision and Pattern Recognition (CVPR) in 1992. For this last conference, Healey, Shafer and Wolff assembled an

extensive collection of representative papers [64, 207, 208]. What progress
has been made over the past few years? What remains to be done? This
chapter reviews research activities since 1988 and closes with a discussion
on the overall success of the field.

8.1 Intrinsic Reflection Images

Pioneering work by Barrow and Tenenbaum has suggested more than a
decade ago to split image data into a set of intrinsic images each of which
is closely related to a single optical scene property [7]. Yet, it was not clear
how to realize the task. Rubin and Richards were the first to demonstrate
how color information can be used to characterize some intrinsic scene
properties [157], followed by Shafer's paper introducing the Dichromatic
Reflection Model and outlining its potential for determining two intrinsic
reflection components [161]. His model formed the basis for the work by
Gershon *et al.* [47], for the work presented in this monograph and for
many recent approaches to separate intrinsic body and surface reflection
components.

Several new approaches to separating body reflection from surface reflection have been proposed and implemented over the last few years. This
section presents and compares these approaches, pointing out which data
sources and what scene knowledge each of them uses.

8.1.1 Separating the Reflection Components

Similar to the work in this text, S.W. Lee *et al.* use color to separate
the reflection components [3, 111]. They analyze color variaton in a color
space similar to IHS (intensity, hue and saturation). When transforming
the original RGB data into IHS coordinates, they discount the illumination color in the image, using color information from a white plate. As a
result, the corrected color image looks as if the illumination is white, and
the highlight clusters are all aligned with the intensity direction (i.e., the
diagonal) of the RGB color space. The IHS axes exhibit a more direct corre-
spondance to body color and to surface reflection, shading, shadowing and
interreflection than the original RGB coordinates. Lee *et al.* exploit these
relationships to segment the color images. They either threshold three separate, one-dimensional histograms of hue, intensity and saturation values
or use a region-based approach based on hue and saturation [3, 111].

In another method called *spectral differencing*, S.W. Lee and Bajcsy exploit *Lambertian consistancy* between several color images from different
views [111, 113]. They point out that Lambertian reflection does not change

when the camera moves. Highlights, on the other hand, do change. As a result, the matte clusters from all images largely coincide in color space. Highlight clusters, on the other hand, change locations. For dielectrics illuminated with white light, the clusters move up and down along the matte cluster, depending on the highlight location in each image and the underlying amount of body reflection. Highlight clusters can thus be detected as the difference between color histograms of image pairs.

Nayar *et al.* exploit the properties of a specialized illumination arrangement [133], according to photometric stereo concepts. Their carefully arranged setup of light sources, known as the *hair dryer*, consists of several light sources placed around a light-diffusing sphere. The object inside the sphere is illuminated by one light source at a time. Nayar *et al.* use the resulting sequence of intensity values at each surface point to compute the surface orientations as well as the relative strengths of Lambertian and specular reflection (which, in their terminology, are equivalent to body and surface reflection – see Section 8.3.2). The method has been named *photometric sampling* because it samples the illumination space uniformly and exploits the photometric properties of light reflection from objects.

Park and Tou also use an extended photometric stereo approach to separate the reflection components and to determine local surface orientations [149]. They observe that both the diffuse reflection component (which they call Lambertian reflection component) and the specular reflection component of a pixel depend on the same surface normal orientation. They use this relationship to formulate a constraint on how a pixel value can be decomposed into two reflection components. Using three intensity images taken from different illuminations directions, as well as measurements of surface albedo, roughness and specularity, they iteratively solve the constraint equation to compute the surface normal and the reflection components at each pixel position.

Ikeuchi and Sato use a range finder that provides them with both range and intensity data [77]. They differentiate the three-dimensional positional data to obtain local surface orientations. They then compute the reflection components and interreflections at every pixel, as well as the light source direction, the albedo, the specular strength and the specular sharpness (inversely related to surface roughness) of an object. Using a reflection model that combines physical and geometric optics to describe surface roughness [135], they distinguish between three reflection components, the *Lambertian component*, the *specular lobe component* and the *specular spike component*. They start out by assuming that the object is perfectly Lambertian. Lambert's law of diffuse reflection then gives them a constraint, for every pixel, on the light source direction and the material albedo value. They estimate the best values by a least square fitting method and use the result to determine, for every pixel, how well it fits the Lambertian assumption. Highlight and interreflection pixels show a significant difference between the measured intensity value and the computed Lambertian term and can thus be detected. Next, Ikeuchi and Sato compare the surface normal at every non-Lambertian pixel with the direction of perfect mirror reflection (which they can compute from the light source direction and the viewing direction). They classify non-Lambertian pixels with surface normals close to the specular direction as *specular pixels* and the other pixels as *interreflection pixels.*

Baribeau *et al.* use range and color data to estimate local reflection properties of objects in an art conservation application [5]. They model light reflection according to the Dichromatic Reflection Model, extended with the concept of the bidirectional reflectance-distribution function (BRDF) [139] and the Torrance-Sparrow model for specular reflection on rough surfaces [188]. They determine average surface roughness and Fresnel reflectance parameters from three selected object areas. They use the parameters to synthesize the surface reflection components. The intrinsic surface reflection image is then subtracted from the color image to generate a body reflection image. Baribeau *et al.* suggest using the color image segmentation algorithm from Chapter 4 to automatically provide sample areas for the parameter estimation process.

Wolff exploits the polarization properties of light reflection [201]. He points out that specular reflection generates highly polarized light whereas diffuse reflection does not. He splits the polarization state of light into a parallel component and a perpendicular component. For specular reflection, the perpendicular component is stronger than the parallel component while both components have equal size for diffuse reflection. Thus, the ratio of the two components is close to 1.0 for diffuse reflection and larger than 1.0 for specular reflection. Wolff uses this criterion to distinguish highlights from matte object areas. In his laboratory, he puts his theory to practice by taking two images of a scene, with a linear polarizer placed in front of his camera. The polarizing filter is oriented into two linearly independent directions for the two images. Wolff points out that although his setup is rather unusual, it is not more difficult than turning a wheel of three color filters in front of a camera. His approach could thus be widely applicable if cameras with built-in polarization detection became available.

Brelstaff and Blake detect highlights in black-and-white images [15]. They observe that most real surfaces have some similarity with pure Lambertian surfaces in areas where they don't show a highlight. Hence they look for non-Lambertian image properties to detect image areas that are likely to be highlights. Two heuristics have been shown to be very effective: 1) an image area can be too bright to be Lambertian, and 2) a peak in shading can be too sharp to be Lambertian. Their results indicate that this detection scheme can detect prominent highlights in a variety of real images.

8.1.2 Using intrinsic reflection images

Once intrinsic reflection images are generated, they can be used in many different ways. Many computer vision algorithms, such as shape-from-shading [69], might perform better on the intrinsic body reflection image than on the original color image since the surface reflection effects have been removed. However, to this date, no studies proving or disproving this claim have been conducted.

Surface reflection provides a strong constraint on the local surface orientation in highlight areas. Several researchers have started using this constraint to determine object shapes. Developing a theory for *specular stereo*, Brelstaff and Blake analyze how highlights change position on objects when viewed from two different locations [12, 16]. The geometric analysis of highlights is not always preceded by a highlight detection step. Healey and Binford, as well as Nayar, Sanderson *et al.* exploit the fact that metals exhibit only surface reflection and that the original image thus is already an intrinsic surface reflection image [58, 137, 158].

Besides using either one of the intrinsic reflection images separately, additional power can be gained from using the images together. Algorithms can use the simplified reflection information in the separate reflection images while exploiting the geometric constraints that the two reflection processes impose on the respective shape processes. Clark and Yuille have presented several methods for fusing these shape constraints [24]. Based on a closed form analytical solution, their system uses several different heuristics for imposing weights on the contribution of the Lambertian versus specular reflection component for estimating local surface orientations under the presence of sensor noise.

The intrinsic reflection images have also been used for material classification. Healey *et al.* exploit the fact that dielectrics have both a body reflection component and a surface reflection component, whereas metals exhibit only surface reflection [61, 62]. Wolff uses the polarization state of light as his criterion for material classification: For a fairly large range of specular angles of incidence, the polarization Fresnel ratio for dielectrics is much larger than for metals [198, 199, 202, 205]. Nayar *et al.* have developed a material classification scheme based on the local reflection components and surface orientations that are determined by their photometric sampling technique [136].

8.2   Color Image Segmentation

Chapter 4 has shown how a careful analysis of local color variation can be used to divide color images into areas of homogeneous body and surface reflection properties. The result is a segmentation along material boundaries which ignores color changes along highlight or shading boundaries.

Several other approaches towards physics-based color image segmentation have recently been proposed. S.W. Lee *et al.* present a fast and simple algorithm that works well under certain scene conditions [3]. They use a white reference card to determine and discount the illumination color. They are then able to distinguish material changes from shading, highlights and interreflection in separate hue and saturation histograms of an IHS-based color space. The one-dimensional histograms can be processed faster than a three-dimensional histogram. In a different physics-based approach, Healey describes how normalized color and irradiance discontinuities can be used for color image segmentation [55, 56, 57]. His system first detects edges in an intensity image and then recursively splits a normalized color image into smaller regions until they don't contain any more edge elements. Shafer *et al.* have extended the work presented in the previous chapters of this text, investigating how interreflections between objects affect color appearance [167].

Other approaches to color image segmentation exploit color properties without considering their relationship to physics-based scene models. Such systems are mentioned here because they present alternate mechanisms for integrating a wealth of color information into more abstract, color based image descriptions. Our own approach performs a well-structured sequence of processing steps that generate increasingly complicated hypotheses (see Figure 4.1). Bhanu *et al.* use a much less structured approach towards color image segmentation. Their system uses a genetic algorithm to achieve adaptive segmentation for variable environmental conditions in outdoor scenes. Using a Phoenix-based recursive region splitting scheme [146, 165], the genetic algorithm searches the hyperspace of segmentation parameters to determine the parameter set that maximizes the segmentation quality criteria [11]. Another example is Daily's use of Markov random fields to segment color images of natural terrain[28], using color differences based on a color metric developed by Healey and Binford [60]. Finally, Westman *et al.* have devised an iterative segmentation scheme that combines a hierarchical connected component analysis with image enhancement techniques using symmetric neighborhood filters [194].

Recently, several systems have successfully used non-physics-based color image and color space analysis in applications, such as road following and object recognition. Crisman uses multi-class Gaussian color models to distinguish between road and off-road pixels in CMU's automatic road follower [26, 27]. Her system describes color variation on and off the road by Gaussian probability functions which are generated from the the mean values and covariance matrices of color clusters in various image regions. The probability distributions partition the color space into piecewise hyperquadric functions. The resulting segmentation of the color space is used to classify the color pixels in the next image which – in turn – generate new color models. Using this two-stage concept, Crisman's system is able to adapt to changing illumination and road conditions.

Swain has developed an approach to object recognition and color image segmentation, called *color indexing*. He observes that object color is a very robust object property and much less susceptible to viewing transformations and occlusion than object shape, yet much faster to compute. He uses a data base of color histograms for a large set of color images and a very fast *histogram intersection* technique to determine whether an object shown in a new image is contained within the model database. In a another approach, he uses the same histograms and *histogram backprojection* to find any of the model objects in a crowded image [174, 175, 176]. Funt and Finlayson have recently extended the color indexing scheme to become more robust under changing illumination colors. They use histograms of color ratios between neighboring color pixels rather than histograms of the color pixels themselves [43]. The techniques work well on flat objects which do not exhibit significant shading variation.

8.3   Reflection and Camera Models

One of the most important characteristics of our work is the fact that it uses a physical reflection model [163]. Yet, the model describes only an approximation to the light reflection processes occurring in real scenes. This section presents two investigations into the validity of the Dichromatic Reflection Model and then reviews research introducing new reflection and camera models, as well as new models which describe color variation in histograms more precisely.

8.3.1 Testing the Dichromatic Reflection Model

Despite the success of our approach, the general question remains: Does the Dichromatic Reflection Model really model the physical reflection processes in the scene? Does it work equally well for all dielectrics, or, if not, where does it succeed or fail? H.-C. Lee *et al.*, as well as Tominaga have recently reported results of such investigations [110, 184].

H.-C. Lee *et al.* have developed the Neutral Interface Reflection Model (NIR) which is based on the bidirectional spectral-reflectance distribution function (BSRDF) [169]. The NIR model is the same as the Dichromatic Reflection Model with $c_s(\lambda)$ constant. They have tested the accuracy of the NIR model on a number of materials and determined that it works well for plastics, plant leaves, painted surfaces, orange peels and some glossy cloth, but not for colored paper and some ceramics.

Tominaga has designed and performed a test suite to validate the Dichromatic Reflection Model itself. For each object, he has measured the spectral power distribution of the reflected light with a spectroradiometer. He has tested whether all measurements from a single material form a two-dimensional plane in the measurement space. He has also checked whether the neutral interface assumption of the Dichromatic Reflection Model is met, i.e., whether the surface reflection component has the same color as the illumination. He reports that the model applies well to plastics (cup, ashtray, box, case, blind), paints (painted metal, door, locker, desk), ceramics (cup, dish), vinyls (tape, cover, sheet), tiles (floor tiles), fruits (apple, lemon), leaves (lemon, kalanchoe, lily) and woods (Japanese cypress), whereas it fails on metals (copper, brass, plating,), cloths (silk, wool, satin) and papers (book, sheet).

8.3.2 Other Reflection Models

In addition to the Dichromatic Reflection Model, several new models of light reflection have emerged during the past few years. Nayar *et al.* model the same physical processes, but describe the geometrical reflection properties more precisely by including a surface roughness parameter[135]. Their model combines a physical optics model [8] for very smooth surfaces with a geometric optics model [188] for rougher surfaces. Within this framework they can distinguish three reflection components: the *Lambertian component* or *diffuse lobe*, the *specular spike*, and the *specular lobe*. When compared to the Dichromatic Reflection Model, the diffuse lobe corresponds to the body reflection component while the specular lobe and specular spike describe surface reflection properties (in contrast to prior use of the term *specular* – see Section 2.1.4), accounting for different surface roughnesses: for very rough surfaces, the specular spike is minimal while the specular

lobe is large. The inverse is true for very smooth surfaces. Surfaces of inter-
mediate roughness may exhibit both a specular lobe and a specular spike.
Other reflection models include interreflection [30, 79, 167], and reflection
from metals [55]. Wolff's reflection model describes the polarization state
of reflected light [206]. New studies revise and extend the commonly used
Lambertian model for diffuse reflection, modeling how subsurface multiple
scattering is altered by Fresnel attenuation upon refraction into air [204].
A survey of various reflection models is given in [177].

8.3.3   Camera Models
Besides modeling the reflection processes in the scene it is essential that
image interpretation algorithms also model camera properties to account
for artifacts due to the imaging process [9, 13, 63, 102, 103, 104, 129, 145].
It has recently become popular to exploit characteristic camera behavior,
such as chromatic aberration and focus, as additional sources of information
[44, 130, 131, 173].

8.3.4   Three-Dimensional Color Space Investigations
The research presented in the previous chapters models and exploits
relationships between optical reflection properties in scenes and their appearances as skewed T's in color histograms. Some recent research has
investigated whether other color spaces, such as CIELAB and IHS, lend
themselves more easily to color image analysis and segmentation than RGB
[3, 183].

Furthermore, several papers have recently investigated color space properties in more depth, considering not only shading and highlights, but also
interreflection between objects, surface roughness [41, 79, 143], and color
cluster changes caused by a moving camera [111, 113]. Novak and Shafer
have analyzed the anatomy of color histograms. They show how three
scene properties – the illumination intensity, the surface roughness, and
the phase angle between camera and light source – are related to the histogram shape. They have devised an algorithm which recovers such scene
properties from the histograms both of noisy synthetic data and of real
images [140, 144].

8.3.5   Spectral Approximations
Beyond exploiting color for image understanding, some research has focussed on the general principles of how to approximate a continuous light
spectrum (an infinite-dimensional measurement space) adequately by a
limited amount of data. Such research activities have investigated several sets of mathematical basis functions, evaluating how well any par-

ticular set of functions is able to represent the reflectance properties of a
broad set of naturally occurring materials under various illumination colors
[3, 32, 60, 67, 107]. Spectral approximations are essential to the theory of
color constancy that is discussed in Section 8.4.3.

8.4   Analyzing Other Optical Phenomena

Many optical phenomena other than shading and highlights coexist in even
moderately complicated scenes. The effects of all are combined in the pixel
values. Intensive investigations have now begun to analyze how individual
phenomena contribute to the image data.

8.4.1   Interreflection

Object interreflection occurs when light is reflected from one object onto
another one. Both the body and the surface reflection components of the
first object can serve as secondary, colored sources of illumination for the
second object, with each component interacting with the second object according to the laws of surface and body reflection. The result are four different combinations of primary and secondary body and surface reflection
processes: body-to-body, body-to-surface, surface-to-body and surface-to-
surface interreflections [3, 142, 167]. The final reflected light is a mixture
of four rather than two colors accounted for in the Dichromatic Reflection
Model: the illumination color, the body colors of both objects, and a new
color which is formed by multiplying the body colors of both objects at
every wavelength. The geometric reflection properties in the interreflection areas are also influenced by the additional light from secondary light
sources, such that simple shape-from-shading approaches render wrong object descriptions. The reflection models become even more complicated
when multiple interreflections are considered in which the light can interact with more than two objects (multiple bounce models). Ray tracing
and radiosity-based approaches in computer graphics model interreflections when rendering realistic synthetic images [35]. Ray tracing methods concentrate on multiple specular interreflections (i.e., surface reflection
from smooth surfaces). Radiosity focusses on diffuse interreflections (i.e.,
body reflection and surface reflection from rough surfaces). The mixed
terms, diffuse-to-specular and specular-to-diffuse, are typically ignored. In
physics-based computer vision, several research projects have now started
to investigate some or all of the four types of interreflections in real images.
Recently, several approaches determine and interpret interreflection in im-

ages by investigating either the color properties or the geometric properties of interreflection [3, 30, 31, 38, 39, 40, 41, 79, 132, 134, 143].

8.4.2 Shadows

Shadows also influence the image data because pixel values in shadow areas are darker than the surrounding lit areas of the scene. Depending on the illumination arrangement, shadows may be completely black or partially illuminated by secondary light sources and ambient light. It has also been pointed out that the color properties of image data from partially shadowed areas may change, due to colored ambient light such as a blue sky [46, 161]. Yet, although shadows may be considered to be an additional problem in the image understanding process, they can also be exploited as an additional source of information about the scene. *Shape-from-shadows* and *shape-from-darkness* approaches relate the size and position of shadow areas to three-dimensional scene properties, such as object shape and the relative position of objects and light sources [1, 53, 54, 72, 78, 81, 114, 115, 166].

8.4.3 Color Constancy

Color constancy is another field of physics-based computer vision research. The color pixel values in images are the product of both the object reflectance properties and the illumination color. Yet, humans usually assign constant color attributes to objects, independently of the illumination. Color constancy research tries to mimic this human ability of discounting the illumination color. It is essential for recognizing objects from their colors.

Several different approaches for achieving color constancy have recently been investigated, such as supervised color constancy using a color chart or a white reference card to normalize the color data [3, 141]. Other methods predetermine the relationship between object colors, illumination colors and observed image intensities for a substantial subset of light spectra and reflectance curves of real-world materials, capturing it in the form of a few basis functions [119, 34, 65] and lookup tables [36, 37]. Other algorithms exploit the highlight color of dielectrics as an indication of the illumination color [33, 108, 185, 186], compute color ratios between neighboring pixels to discount the illumination [43], or use color variation between two images of the same scene, taken under different illumination, to estimate the object and illumination colors [189].

In its inverse formulation, color constancy concept can also be used to design illumination environments according to specific characteristics. Vriesenga *et al.* use such a formalism to obtain maximal contrast between

the colors of potato leaves and stems or between a collection of color patches
[192].

8.4.4 Polarization

Wolff's research on polarized light reflection is a good example demonstrating how models of specific light reflection properties can be used to
relate sensor data to scene properties. As described in Section 8.1.1, the
polarization properties of light reflection can serve as a means for distinguishing between specular reflection and diffuse reflection [201, 203]. Such
information can be used to classify the objects in the scene into metals
and dielectrics [198, 199, 202, 205] and to label image edges as occluding,
specular, or albedo (physical) edges [14].

8.4.5 Shape from Shading in Image Sequences

Two classical fields of physics-based *shape-from-X* approaches have recently started to converge: the analysis of image sequences and the analysis
of shading. In the past, these two approaches were considered orthogonal:
optical flow, stereo and motion algorithms typically use purely geometric
models when finding corresponding image features in several images. They
work best for highly textured objects because texture generates many intensity edges that can be matched and tracked. Shape-from-shading, on
the other hand, works best on untextured objects. It generally analyzes the
photometric aspects of the image formation process in a single image, ignoring the wealth of additional information that can be obtained from considering a sequence of images under changing image formation conditions.
An early exception to this practice is photometric stereo [74, 209] which
systematically exploits photometric changes in sequences of images taken
under a variety of different, controlled lighting arrangements. Recently,
photometric stereo approaches have started analyzing specular reflection
in addition to diffuse reflection and are also able to estimate changing
albedos [170, 178, 179, 180]. As an alternative to moving light sources,
S.W. Lee and Bajcsy suggest exploiting color changes caused by a moving
camera (see Section 8.1.1) [111, 113].

In the *shape-from-motion* and *shape-from-stereo* field, recent research activities have started considering photometric changes on moving objects
in addition to the geometric changes. Pentland and others point out that
the photometric changes of a point on a moving object can confound the
geometric changes, leading to incorrect matches during the search for corresponding points in consecutive images, and thus to wrong shape estimates
[151, 168]. A few research activities have now endeavoured to extend the
traditional stereo or motion analysis approaches with reflection models and

shape-from-shading methods [52, 151, 168, 210, 200] or even an analysis of specular reflection [12, 16, 127].

8.4.6   Other Work

Besides combining shape-from-shading approaches with object motion analysis, research considering other scene properties in conjunction with a shading analysis has emerged. Among such efforts are attempts to include the shape-from-texture paradigm [21, 22, 181], to combine shape-from-shading with range data analysis [77, 120], and estimation of the direction of one or more light sources while estimating the object shape [127, 211, 213]. Some research turns the light source estimation problem into a camera and light source preplanning problem [23, 105, 192, 212], an approach known as *active vision* [2]. Furthermore, investigations towards discounting the effects of object color (albedo) on shading have recently been presented [42].

8.5   Are We There Yet?

As stated at the beginning of this chapter, physics-based computer vision has gained much visibility over the past few years. The research community has grown rapidly and produced many new exciting results. Yet, many questions remain: How much progress have we made so far? What are the overall goals? Have, will, or can we reach them? Will physics-based computer vision be able to replace or enhance traditional computer vision methods in useful applications?

Physics-based computer vision is a very fascinating field that thrills new-comers due to its relationship to the physical reality of the world. Intu-itively, it seems to be the right way to analyze images. Yet, many people are very quickly discouraged by the number of limiting assumptions that have to be made. So far, most algorithms have only demonstrated success on very few images which were typically taken in a carefully arranged setup (e.g., in the Calibrated Imaging Lab at CMU) or were synthetically gener-ated. Considering, for example, applications in which a robot is expected to navigate autonomously outdoors, in an office, or in a kitchen, we notice that current navigation systems seem to do better without physics-based vision than with it. What's missing? Many scene properties of typical outdoor, office or kitchen scenes have not been considered yet in physics-based com-puter vision – and especially not in combination with one another. Here are a few examples: Realistic scenes contain objects with smoothly changing albedo, such as paintings, and objects with regular and irregular textures (both in terms of surface roughness and surface coloring,) such as carpets, grass, wood grain and walls. Some objects, such as windows and doors, are transparent or translucent, and some rooms have mirrors – not to mention cigarette smoke and fog. Furthermore, we are generally surrounded by many light sources, ambient light (e.g., a blue sky), with shadows, interreflections, etc. All these different physical scene properties with many different parameters contribute to the image formation process. Physicists and the computer graphics community work hard, trying to determine appropriate reflection models and parameter values (many of which describe specific material properties) so that they can generate realistic synthetic images. It is even harder for computer vision algorithms to invert the image formation process and try to estimate this wealth of parameters and the intricate relationships between illuminations, object reflection properties and object positions from images. Most likely, some limiting assumptions will always have to be imposed on the scenes. Furthermore, estimating even a small number of physical scene properties tends to be prohibitively time-consuming.

Should we thus give up? I do not think so. Physics-based computer vision is absolutely essential, if computer vision wants to succeed in the long run. An analogy: it took a leap in technology to fly to the moon. Trying to get there with a ladder just wasn't good enough. Of course, ladders have been good for picking apples from trees long before the first space rocket ever left the ground. And even today, ladders seem to be more appropriate for this task. Yet, despite such long-established advantages of ladders over space rockets, incrementally extending ladders by more and more steps just won't get us to the moon. More seriously: traditional computer vision has provided a large selection of techniques that tend to work moderately robustly over a range of images. But they fail in rather unpredictable ways when a physical scene property, such as a highlight, has a more than negligible effect on the image data. Such failures are nearly impossible to correct in traditional algorithms because the algorithms have not been designed to account for the underlying physical causes. Typical examples of such problems have been pointed out in chapters 1 and 6, in a comparison between the physics-based approach and Phoenix, a more traditional segmentation system.

Such issues can become critical in some application areas, thus demanding a physics-based approach. Lee and Bajcsy report that they are using their highlight detection system to detect highlight lines on cars in a complicated illumination environment, providing the results as input to a CAD-model for car design [112]. Shafer and Johnston-Feller suggest replacing convential *gloss traps* [87] in spectrophotometers with an electronic highlight removal system [164]. Brunner *et al.* exploit color information for automatic wood inspection [17, 121]. They point out what problems specularities cause for automatic wood inspection in the forest industry. Wood color is seriously altered by specular reflection and can lead to misclassifications of wood samples. The amount of specular reflection depends on the fiber angle which cannot always be carefully controlled [18]. Finally, Nayar's shape from focus system [131] is built according to the needs of an industrial sponsor. This system inspects a sequence of light microscope images taken under varying focal lengths. It cuts and pastes together image regions from the sequence that have the highest high frequency content (i.e., sharpest edges), relating the focal length under which those regions where viewed to object distance.

Even if physics-based computer vision could be useful, how are we going to use it in real applications? An equal and simultaneous treatment of all possible physical scene properties might not be achievable. The research presented in this text suggests a more pragmatic approach. Based on the observation that most pixels in most images are rather simple to interpret and that only a few pixels cause most of the confusion, we have distinguished between several pixel classes: very dark pixels, matte pixels, highlight pixels, bloomed and clipped pixels. The segmentation algorithm in chapter 4 (see Figure 4.1) operates incrementally, starting by assuming that all pixels are simply very dark or matte, thus identifying those pixels with the simplest color properties. It then refines the segmentation by taking the more complicated color properties of highlight pixels into account. Nayar et al. [134] are using a similar paradigm to correctly estimate object shape from image areas in which shading was altered by interreflections: They also start out using simplifying assumptions (the traditional shape-from-shading approach) and then improve the solution iteratively by considering interreflection effects in more and more detail.

When many different physical scene properties have to be considered, some in combination with others, this may lead to a rather complex physics-based image interpretation system. Existing vision systems [50] have not based their interpretation primitives on physical scene properties. From looking at past experience, it becomes clear that such system development can only be successful in a suitable software development environment which might be called a *Vision Programmer's Workbench* in which different independent modules focus on specific aspects of the physical processes (e.g., on the photometric or geometric highlight reflection properties of dielectrics, or on interreflections, or shadows), while communicating via some medium, such as a blackboard. Several systems to help researchers develop and visualize image analysis techniques are beginning to emerge [49, 94, 95, 96, 101, 116, 123, 124, 125, 128, 153, 154, 172, 191, 195].

Instead of trying to solve the general vision problem, application-specific simplifying assumptions may make the difference between a basic (and slow) research tool and a practical, economically viable system. Many industrial applications may be engineered according to limiting capabilities of some technology, such as Wolff's polarization work [203], Nayar's hair dryer [133], photometric stereo approaches [74, 209], and supervised color constancy [3, 141]. The image interpretation task can then be simplified tremendously because some amount of information can be provided. Conversely, new physics-based approaches can be developed with specific applications in mind[131, 192], as discussed above.

So: are we there yet? No, we have barely started. It will be a long way before physics-based computer vision is useful to a broad area of applications, and even longer before it can be used real-time on-board a navigating robot system. But this text, the many research activities currently under way, and the emerging applications are a very encouraging start. Computer vision will not solve the general vision problem without physics-based reasoning.

A

Derivation of the 50%-Heuristic

Figure A.1 shows the illumination geometry that is the basis for deriving the 50%-heuristic. It shows a spherical object with radius r which is viewed and illuminated under perspective projection . The following analysis assumes that the sphere is illuminated by a point light source at some distance d from the object center. It further assumes that the camera is positioned at the same distance d from the object center and that an arbitrary phase angle g exists between the viewing and the illumination direction.

The distance d determines the size of the cone that illuminates the sphere (as described by angle γ), as well as the range of phase angles $[-g_{max}, g_{max}]$ under which the camera sees a part of the illuminated area:

$$\gamma = \arctan \frac{r}{\sqrt{d^2 - r^2}} \qquad (A.1)$$

$$g_{max} = 180^\circ - 2\gamma = 2 \arctan(\frac{\sqrt{d^2 - r^2}}{r}). \qquad (A.2)$$

The range of phase angles is indicated in Figure A.1 as a sequence of camera positions, arranged on a circular segment.

Figure A.2 describes how the highlight and its underlying body reflection component m_{bH} are related to phase angle g. Since camera and light source are positioned at the same distance d from the object center, the maximal

129

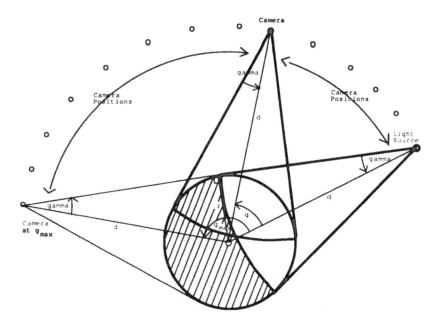

Figure A.1. Assumed illumination geometry.

surface reflection component occurs where the surface normal, n_H, bisects the phase angle. Under perspective projection, the illumination angle i_H at the highlight depends on the position of the highlight which, in turn, depends on g. Using the law of sines and the law of cosines, i_H can be described as

$$i_H = \arcsin \frac{d \sin(g/2)}{\sqrt{d^2 + r^2 - 2dr \cos(g/2)}}. \tag{A.3}$$

Assuming Lambertian body reflection, $m_b(i, e, g) = \cos i$, the underlying body reflection component, m_{bH}, at the highlight is then given by

$$m_{bH} = \cos(i_H) = \sqrt{1 - \frac{d^2 \sin^2(g/2)}{d^2 + r^2 - 2dr \cos(g/2)}}. \tag{A.4}$$

According to equations A.3 and A.4, m_{bH} approaches 1 when g goes to $0°$, confirming the informal description at the beginning of Section 2.3.2 that the spectral cluster looks like a skewed L when camera and light source

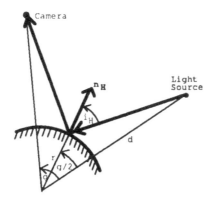

Figure A.2. Photometric angles at highlight.

are close. When g approaches g_{max}, on the other hand, i_h approaches $90°$, by the definition of g_{max} (see Figure A.1). The underlying body reflection component then decreases to 0. This shows that the body reflection under the highlight m_{bH} can vary between 0% and 100% of the maximum body reflection on the object. However, the point with globally maximal body reflection, m_{bMax}, is not always visible from the camera. If it is occluded, the matte line in spectral histogram does not extend entirely to $m_{bMax}c_b(\lambda)$ but only to the point representing the brightest visible point, $m_{bLocalMax}c_b(\lambda)$.

Figure A.3 displays the illumination geometry for the brightest matte point that is visible from the camera. The surface normal, $n_{LocalMax}$, at this object point is determined by the angle α, which is given as

$$\alpha = \max(g - (90° - \gamma), 0) = \begin{cases} g - (g_{max}/2) & \text{if } g \geq g_{max}/2 \\ 0 & \text{otherwise.} \end{cases} \quad (A.5)$$

Following the derivation of equations A.3 and A.4, the angle between the illumination vector and the surface normal at the local maximum and the amount of body reflection at that point are given by equations A.6 and A.6:

$$i_{LocalMax} = \arcsin \frac{d \sin \alpha}{\sqrt{d^2 + r^2 - 2dr \cos \alpha}}, \quad (A.6)$$

$$m_{bLocalMax} = \cos(i_{LocalMax}) = \sqrt{1 - \frac{d^2 \sin^2 \alpha}{d^2 + r^2 - 2dr \cos \alpha}}. \quad (A.7)$$

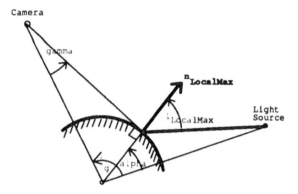

Figure A.3. Brightest matte point that is visible from the camera.

The starting point of the highlight line from the matte line can now be described relative to the length of the matte line. The ratio q of body reflection at the highlight in relation to the maximally visible amount of body reflection is expressed as

$$q = \frac{m_{bH}}{m_{bLocalMax}}. \qquad (A.8)$$

B

Tables of Illumination Geometries

Table B.1. Body reflection components and quotient q for $d = 1.01r$.

$r = 1.00$, $d = 1.01$, $\gamma = 81.93°$, $g_{max} = 16.14°$

g	$i_{LocalMax}$	$m_{bLocalMax}$	i_H	m_{bH}	q
0.00	0.00	1.00	0.00	1.00	1.00
2.00	0.00	1.00	60.81	0.49	0.49
4.00	0.00	1.00	75.09	0.26	0.26
6.00	0.00	1.00	80.74	0.16	0.16
8.00	0.00	1.00	83.89	0.11	0.11
10.00	74.52	0.27	86.00	0.07	0.26
12.00	83.72	0.11	87.58	0.04	0.39
14.00	87.48	0.04	88.85	0.02	0.46
16.00	89.86	0.00	89.93	0.00	0.50
16.13	89.99	0.00	89.99	0.00	0.50

Table B.2. Body reflection components and quotient q for $d = 1.05r$.

$r = 1.00$, $d = 1.05$, $\gamma = 72.25°$, $g_{max} = 35.51°$

g	$i_{LocalMax}$	$m_{bLocalMax}$	i_H	m_{bH}	q
0.00	0.00	1.00	0.00	1.00	1.00
2.00	0.00	1.00	20.19	0.94	0.94
4.00	0.00	1.00	36.39	0.80	0.80
6.00	0.00	1.00	48.53	0.66	0.66
8.00	0.00	1.00	57.07	0.54	0.54
10.00	0.00	1.00	63.31	0.45	0.45
12.00	0.00	1.00	68.04	0.37	0.38
14.00	0.00	1.00	71.76	0.31	0.31
16.00	0.00	1.00	74.77	0.26	0.26
18.00	5.18	1.00	77.28	0.22	0.22
20.00	39.93	0.77	79.42	0.18	0.24
22.00	58.79	0.52	81.29	0.15	0.29
24.00	69.04	0.36	82.94	0.12	0.34
26.00	75.43	0.25	84.42	0.10	0.39
28.00	79.91	0.18	85.76	0.07	0.42
30.00	83.32	0.12	87.00	0.05	0.45
32.00	86.08	0.07	88.15	0.03	0.47
34.00	88.43	0.03	89.23	0.01	0.49
35.50	89.99	0.00	89.99	0.00	0.50

Table B.3. Body reflection components and quotient q for $d = 1.1r$.

$r = 1.00$, $d = 1.10$, $\gamma = 65.38°$, $g_{max} = 49.24°$

g	$i_{LocalMax}$	$m_{bLocalMax}$	i_H	m_{bH}	q
0.00	0.00	1.00	0.00	1.00	1.00
5.00	0.00	1.00	25.87	0.90	0.90
10.00	0.00	1.00	45.02	0.71	0.71
15.00	0.00	1.00	57.75	0.53	0.53
20.00	0.00	1.00	66.44	0.40	0.40
25.00	4.17	1.00	72.75	0.30	0.30
30.00	47.31	0.68	77.61	0.21	0.32
35.00	67.52	0.38	81.56	0.15	0.38
40.00	78.26	0.20	84.89	0.09	0.44
45.00	85.35	0.08	87.79	0.04	0.48
49.23	89.99	0.00	89.99	0.00	0.50

Table B.4. Body reflection components and quotient q for $d = 1.2r$.
$r = 1.00$, $d = 1.20$, $\gamma = 56.44°$, $g_{max} = 67.11°$

g	$i_{LocalMax}$	$m_{bLocalMax}$	i_H	m_{bH}	q
0.00	0.00	1.00	0.00	1.00	1.00
5.00	0.00	1.00	14.75	1.00	0.97
10.00	0.00	1.00	28.15	0.88	0.88
15.00	0.00	1.00	39.54	0.77	0.77
20.00	0.00	1.00	48.90	0.66	0.66
25.00	0.00	1.00	56.55	0.55	0.55
30.00	0.00	1.00	62.87	0.46	0.46
35.00	8.61	1.00	68.18	0.37	0.38
40.00	34.98	0.82	72.73	0.30	0.36
45.00	53.50	0.59	76.69	0.23	0.39
50.00	66.04	0.41	80.20	0.17	0.42
55.00	75.07	0.26	83.37	0.12	0.45
60.00	82.07	0.14	86.26	0.07	0.47
65.00	87.82	0.04	88.93	0.02	0.49
67.10	89.99	0.00	89.99	0.00	0.50

Table B.5. Body reflection components and quotient q for $d = 1.5r$.
$r = 1.00$, $d = 1.50$, $\gamma = 41.81°$, $g_{max} = 96.38°$

g	$i_{LocalMax}$	$m_{bLocalMax}$	i_H	m_{bH}	q
0.00	0.00	1.00	0.00	1.00	1.00
10.00	0.00	1.00	14.81	0.97	0.97
20.00	0.00	1.00	28.63	0.88	0.88
30.00	0.00	1.00	40.86	0.76	0.76
40.00	0.00	1.00	51.40	0.62	0.62
50.00	5.42	1.00	60.45	0.49	0.50
60.00	33.25	0.84	68.26	0.37	0.44
70.00	54.83	0.58	75.11	0.26	0.45
80.00	70.84	0.33	81.21	0.15	0.47
90.00	83.27	0.12	86.73	0.06	0.49
96.37	89.99	0.00	89.99	0.00	0.50

Table B.6. Body reflection components and quotient q for $d = 2r$.
$r = 1.00$, $d = 2.00$, $\gamma = 30.00°$, $g_{max} = 120.00°$

g	$i_{LocalMax}$	$m_{bLocalMax}$	i_H	m_{bH}	q
0.00	0.00	1.00	0.00	1.00	1.00
10.00	0.00	1.00	9.96	0.99	0.98
20.00	0.00	1.00	19.71	0.94	0.94
30.00	0.00	1.00	29.05	0.87	0.87
40.00	0.00	1.00	37.88	0.79	0.79
50.00	0.00	1.00	46.13	0.69	0.69
60.00	0.00	1.00	53.79	0.59	0.59
70.00	19.71	0.94	60.91	0.49	0.52
80.00	37.88	0.79	67.52	0.38	0.48
90.00	53.79	0.59	73.68	0.28	0.48
100.00	67.52	0.38	79.44	0.18	0.48
110.00	79.44	0.18	84.87	0.09	0.48
119.99	89.99	0.00	89.99	0.00	0.50

Table B.7. Body reflection components and quotient q for $d = 5r$.
$r = 1.00$, $d = 5.00$, $\gamma = 11.54°$, $g_{max} = 156.93°$

g	$i_{LocalMax}$	$m_{bLocalMax}$	i_H	m_{bH}	q
0.00	0.00	1.00	0.00	1.00	1.00
10.00	0.00	1.00	6.25	0.99	0.99
20.00	0.00	1.00	12.48	0.98	0.95
30.00	0.00	1.00	18.67	0.95	0.95
40.00	0.00	1.00	24.81	0.91	0.91
50.00	0.00	1.00	30.89	0.86	0.86
60.00	0.00	1.00	36.90	0.80	0.80
70.00	0.00	1.00	42.81	0.73	0.73
80.00	1.92	1.00	48.63	0.66	0.66
90.00	14.39	0.97	54.35	0.58	0.60
100.00	26.69	0.89	59.97	0.50	0.56
110.00	38.72	0.78	65.48	0.41	0.53
120.00	50.40	0.64	70.89	0.33	0.51
130.00	61.68	0.47	76.20	0.24	0.50
140.00	72.54	0.60	81.41	0.15	0.50
150.00	82.99	0.12	86.52	0.06	0.50
156.92	89.99	0.00	89.99	0.00	0.50

Table B.8. Body reflection components and quotient q for $d = \infty$.

$r = 1.00$, $d = 5.00$, $\gamma = 11.54°$, $g_{max} = 156.93°$

g	$i_{LocalMax}$	$m_{bLocalMax}$	i_H	m_{bH}	q
0.00	0.00	1.00	0.00	1.00	1.00
10.00	0.00	1.00	5.00	1.00	1.00
20.00	0.00	1.00	10.00	0.98	0.98
30.00	0.00	1.00	15.00	0.97	0.97
40.00	0.00	1.00	20.00	0.94	0.94
50.00	0.00	1.00	25.00	0.91	0.91
60.00	0.00	1.00	30.00	0.87	0.87
70.00	0.00	1.00	35.00	0.82	0.82
80.00	0.00	1.00	40.00	0.77	0.76
90.00	0.00	1.00	45.00	0.71	0.71
100.00	10.00	0.98	50.00	0.64	0.65
110.00	20.00	0.94	55.00	0.57	0.61
120.00	30.00	0.87	60.00	0.50	0.58
130.00	40.00	0.77	65.00	0.42	0.55
140.00	50.00	0.64	70.00	0.34	0.53
150.00	60.00	0.50	75.00	0.26	0.52
160.00	70.00	0.34	80.00	0.1736	0.51
170.00	80.00	0.17	85.00	0.0872	0.50
179.99	89.99	0.00	89.99	0.0001	0.50

C

Illumination Geometry for $d = \infty$

When the light source is infinitely far away ($d = \infty$), incident light rays at all object points are parallel to one another such that the laws of orthographic projection apply. The illuminating cone then collapses into a line ($\gamma = 0°$) while the phase angles $\|g\|$ range between $0°$ and $180°$ since $g_{max} = 180°$. In this case, the computation of the illumination angles at the highlight and at the brightest visible point simplifies to

$$i_H = g/2 i_{LocalMax} = \alpha = \begin{cases} 0 & \text{if } g \leq 90° \\ g - 90° & \text{if } g \geq 90°. \end{cases} \tag{C.1}$$

The amount of body reflection at these points is then determined by

$$m_{bH} = \cos(g/2) m_{bLocalMax} = \begin{cases} \cos(0) = 1 & \text{if } g \leq 90° \\ \cos(g - 90°) = \sin(g) & \text{if } g \geq 90°. \end{cases} \tag{C.2}$$

Accordingly, the quotient q of the two body reflection components is given by

$$q_\infty(g) = \begin{cases} \cos(g/2) & \text{if } g \leq 90° \\ \cos(g/2)/\sin(g) = 1/(2\sin(g/2)) & \text{if } g \geq 90°. \end{cases} \tag{C.3}$$

References

[1] M. Adjouadi. Image analysis of shadows, depressions and upright objects in the interpretation of real-world scenes. In *IEEE International Conference on Pattern Recognition (ICPR'86)*, pages 73–76. IEEE, 1986.

[2] J. Aloimonos, I. Weiss, and A. Mandyopadhyay. Active vision. In *Proceedings of the First International Conference on Computer Vision (ICCV)*, pages 35–54, London, June 1987. IEEE.

[3] R. Bajcsy, S.W. Lee, and A. Leonardis. Color image segmentation with detection of highlights and local illumination induced by inter-reflections. In *IEEE 10th International Conference on Pattern Recognition (ICPR'90)*, pages 785–790, Atlantic City, NJ, June 1990. IEEE.

[4] D.H. Ballard and C.M. Brown. *Computer Vision.* Prentice Hall, Inc., Englewood Cliffs, NJ, 1982.

[5] R. Baribeau, M. Rioux, and G. Godin. Color reflectance modeling using a polychromatic laser range finder. *IEEE Trans. on Pattern Analysis and Machine Intelligence (PAMI)*, 14(2):263–269, February 1992.

[6] S.T. Barnard and M.A. Fischler. Computational stereo. *Computing Surveys*, 14(4):553–572, December 1982.

[7] H.G. Barrow and J.M. Tenenbaum. Recovering intrinsic scene characteristics from images. In A.R. Hanson and E.M. Riseman, editors, *Computer Vision Systems*, pages 3–26. Academic Press, New York, 1978.

[8] P. Beckmann and A. Spizzichino. *The Scattering of Electromagnetic Waves from Rough Surfaces*, volume 15. MacMillan, New York, 1963. Pages 1-33, 70-96.

[9] H.A. Beyer. Accurate calibration of CCD-cameras. In *IEEE Conference on Computer Vision and Pattern Recognition (CVPR'92)*, pages 96–101, Champaign, IL, June 1992. IEEE.

[10] J.D.E. Beynon and D.R. Lamb (eds.). *Charge-Coupled Devices and Their Application.* McGraw-Hill, London, 1980.

[11] B. Bhanu, J. Ming, and S. Lee. Closed-loop adaptive image segmentation. In *IEEE Conference on Computer Vision and Pattern Recognition (CVPR'91)*, pages 734–735, Lahaina, Maui, Hawaii, June 1991. IEEE.

[12] A. Blake and G. Brelstaff. Geometry from specularities. In *Proceedings of the second International Conference on Computer Vision (ICCV)*, pages 394–403, Tampa, Florida, Dec 1988. IEEE.

[13] T.E. Boult and G. Wolberg. Correcting chromatic aberrations using image warping. In L.S. Bauman, editor, *DARPA-Image Understanding (IUS) workshop*, pages 363–378, San Diego, CA, January 1992. Morgan Kaufmann.

[14] T.E. Boult and L.B. Wolff. Physically-based edge labelling. In *IEEE Conference on Computer Vision and Pattern Recognition (CVPR'91)*, pages 656–662, Lahaina, Maui, Hawaii, June 1991. IEEE.

[15] G. Brelstaff and A. Blake. Detecting specular reflections using Lambertian constraints. In *Proceedings of the second International Conference on Computer Vision (ICCV)*, pages 297–302, Tampa, Florida, Dec 1988. IEEE.

[16] G.J. Brelstaff. *Inferring surface shape from specular reflections.* PhD thesis, Dept. Computer Science, University of Edinburgh, 1989.

[17] C.C. Brunner, A.G. Maristany, D.A. Butler, D. VanLeeuwen, and J.W. Funck. An evaluation of color spaces for detecting defects in Douglas-fir veneer. *Wood and Fiber Science*, 22(4), 1992.

[18] C.C. Brunner, G.B. Shaw, D.A. Butler, and J.W. Funck. Using color in machine vision systems for wood processing. *Wood and Fiber Science*, 22(4), 1990.

[19] W. Budde. *Physical Detectors of Optical Radiation*, volume 4 of *Optical Radiation Measurements*. Academic Press, New York, 1983.

[20] J. Canny. A computational approach to edge detection. *IEEE Trans. on Pattern Analysis and Machine Intelligence (PAMI)*, 8(6):679–698, November 1986.

[21] Y. Choe and R.L. Kashyap. Shape from textured and shaded surface. In *IEEE 10th International Conference on Pattern Recognition (ICPR'90)*, pages 294–296, Atlantic City, NJ, June 1990. IEEE.

[22] Y. Choe and R.L. Kashyap. 3-D shape from a shaded and textural surface image. *IEEE Trans. on Pattern Analysis and Machine Intelligence (PAMI)*, 13(9):907–919, Sep 1991.

[23] J.J. Clark. Active photometric stereo. In *IEEE Conference on Computer Vision and Pattern Recognition (CVPR'92)*, pages 29–34, Champaign, IL, June 1992. IEEE.

[24] J.J. Clark and A.L. Yuille. Shape from shading via the fusion of specular and Lambertian image components. In *IEEE 10th International Conference on Pattern Recognition (ICPR'90)*, pages 88–92, Atlantic City, NJ, June 1990. IEEE.

[25] R.L. Cook and K.E. Torrance. A reflectance model for computer graphics. *ACM Transactions on Graphics*, 1(1):7–24, January 1982. Also published in Computer Graphics 15(3), SIGGRAPH'81.

[26] J. Crisman and C. Thorpe. Color vision for road following. In C. Thorpe, editor, *Vision and Navigation: The Carnegie Mellon Navlab*. Kluwer Academic Publishers, 1990.

[27] J.D. Crisman. *Color Vision for the Detection of Unstructured Roads and Intersections*. PhD thesis, Department of Electrical and Computer Engineering, Carnegie Mellon, May 1990.

[28] M.J. Daily. Color image segmentation using Markov random fields. In L.S. Bauman, editor, *DARPA-Image Understanding (IUS) workshop*, pages 552–562. Morgan Kaufmann, May 1989.

[29] L. Dreschler and H.-H. Nagel. Volumetric model and 3D trajectory of a moving car derived from monocular tv frame sequences of a street scene. *Computer Graphics and Image Processing*, 20:199–228, 1982.

[30] M.S. Drew and B.V. Funt. Calculating surface reflectance using a single-bounce model of mutual reflection. In *Proceedings of the third International Conference on Computer Vision (ICCV)*, pages 394–399, Osaka, Japan, Dec 1990. IEEE.

[31] M.S. Drew and B.V. Funt. Variational approach to mutual illumination in color images. Technical Report CSS/LCCR TR 90-21, Simon Fraser University, Centre For Systems Science LCCR, Burnaby, B.C., Canada V5A 1 S6, Dec 1990.

[32] M.S. Drew and B.V. Funt. Natural metamers. Technical Report CSS/LCCR TR 90-05, Simon Fraser University, Centre For Systems Science LCCR, Burnaby, B.C., Canada V5A 1 S6, Feb 1991.

[33] M. D'Zmura and P. Lennie. Mechanisms of color constancy. *Journal of the Optical Society of America A (JOSA A)*, 3(10):1662–1672, October 1986.

[34] G.D. Finlayson, M.S. Drew, and B.V. Funt. Spectral sharpening: Sensor transformations for improved color constancy. Technical Report CSS/LCCR TR 91-14, Simon Fraser University, Centre For Systems Science LCCR, Burnaby, B.C., Canada V5A 1 S6, Dec 1991.

[35] J.D. Foley, A. van Dam, S. K. Feiner, and J.F. Hughes. *Computer Graphics, Principles and Practice, 2. Edition*. Addison-Wesley Publishing Company, Reading, MA, Menlo Park, CA, New York, Don Mills, Ontario, Wokinghame, England, Amsterdam, Bonn, Sydney, Singapore, Tokyo, Madrid, San Juan, 1989.

[36] D.A. Forsyth. A novel approach to colour constancy. In *Proceedings of the second International Conference on Computer Vision (ICCV)*, pages 9–17, Tampa, Florida, Dec 1988. IEEE.

[37] D.A. Forsyth. A novel algorithm for colour constancy. *International Journal on Computer Vision (IJCV)*, 5(1):5–36, Aug 1990.

[38] D.A. Forsyth and A. Zisserman. Mutual illumination. In *IEEE Conference on Computer Vision and Pattern Recognition (CVPR'89)*, pages 466–473, San Diego, CA, June 1989. IEEE.

[39] D.A. Forsyth and A. Zisserman. Shape from shading in the light of mutual illumination. *Image Vision Computing*, 8:42–49, 1990.

[40] D.A. Forsyth and A. Zisserman. Reflections on shading. *IEEE Trans. on Pattern Analysis and Machine Intelligence (PAMI)*, 13(7):671–679, July 1991.

[41] B.V. Funt and M.S. Drew. Color space analysis of mutual illumination. Technical Report CSS/LCCR TR 91-03, Simon Fraser University, Centre For Systems Science LCCR, Burnaby, B.C., Canada V5A 1 S6, Apr 1991.

[42] B.V. Funt, M.S. Drew, and M. Brockington. Recovering shading from color images. In *Proceedings of the second European Conference on Computer Vision (ECCV)*, pages 124–132, Santa Margherita Ligure, Italy, May 1992. Springer Verlag. Also available as technical report CSS/LCCR TR91-08, Oct. 1991.

[43] B.V. Funt and G.D. Finlayson. Color constant color indexing. Technical Report CSS/LCCR TR 91-09, Simon Fraser University, Centre For Systems Science LCCR, Burnaby, B.C., Canada V5A 1 S6, Oct 1991.

[44] B.V. Funt and J. Ho. Color from black and white. In *Proceedings of the second International Conference on Computer Vision (ICCV)*, pages 2–8, Tampa, Florida, Dec 1988. IEEE.

[45] R. Gershon. *The Use of Color in Computational Vision*. PhD thesis, Department of Computer Science, University of Toronto, 1987.

[46] R. Gershon, A.D. Jepson, and J.K. Tsotsos. Ambient illumination and the determination of material changes. *Journal of the Optical Society of America A (JOSA A)*, 3:1700–1707, October 1986.

[47] R. Gershon, A.D. Jepson, and J.K. Tsotsos. Highlight identification using chromatic information. In *Proceedings of the First International Conference on Computer Vision (ICCV)*, pages 161–171, London, June 1987. IEEE.

[48] H. Grassmann. On the theory of compound colors. *Phil. Mag.*, April 1854.

[49] A.J. Hanson and L.H. Quam. Overview of the SRI Cartographic Modeling Environment. In L.S. Bauman, editor, *DARPA-Image Understanding (IUS) workshop*, pages 576–582. Morgan Kaufmann, April 1988.

[50] A.R. Hanson and E.M. Riseman. Visions: A computer system for interpreting scenes. In A.R. Hanson and E.M. Riseman, editors, *Computer Vision Systems*, pages 303–333. Academic Press, New York, 1978.

[51] H.H. Harman. *Modern Factor Analysis, second edition.* University of Chicago Press, Chicago and London, 1967.

[52] K. Hartt and M. Carlotto. A method for shape-from-shading using multiple images acquired under different viewing and lighting conditions. In *IEEE Conference on Computer Vision and Pattern Recognition (CVPR'89)*, pages 53–60, San Diego, CA, June 1989. IEEE.

[53] M. Hatzitheodorou. The derivation of 3-D surface shape from shadows. In L.S. Bauman, editor, *DARPA-Image Understanding (IUS) workshop*, pages 1012–1020. Morgan Kaufmann, May 1989.

[54] M. Hatzitheodorou and J. Kender. An optimal algorithm for the derivation of shape from darkness. In *IEEE Conference on Computer Vision and Pattern Recognition (CVPR'88)*, Ann Arbor, MI, June 1988. IEEE.

[55] G. Healey. A color reflectance model and its use for segmentation. In *Proceedings of the second International Conference on Computer Vision (ICCV)*, pages 460–466, Tampa, Florida, Dec 1988. IEEE.

[56] G. Healey. A parallel color algorithm for segmenting images of 3-D scenes. In L.S. Bauman, editor, *DARPA-Image Understanding (IUS) workshop*, pages 1038–1042. Morgan Kaufmann, May 1989.

[57] G. Healey. Using color for geometry insensitive segmentation. *Journal of the Optical Society of America A (JOSA A)*, 6(6):920–937, June 1989.

[58] G. Healey and T.O. Binford. Local shape from specularity. In *Proceedings of the First International Conference on Computer Vision (ICCV)*, pages 151–161, London, June 1987. IEEE.

[59] G. Healey and T.O. Binford. The role and use of color in a general vision system. In L.S. Bauman, editor, *DARPA-Image Understanding (IUS) workshop*, pages 599–613, Los Angeles, CA, February 1987. Morgan Kaufmann.

[60] G. Healey and T.O. Binford. A color metric for computer vision. In L.S. Bauman, editor, *DARPA-Image Understanding (IUS) workshop*, pages 854–861, Cambridge, MA, Apr 1988. Morgan Kaufmann.

[61] G. Healey and T.O. Binford. Predicting material classes. In *Proc. of the DARPA-Image Understanding Workshop*, pages 1140–1146, Boston, MA, April 6-8 1988. Morgan Kaufmann.

[62] G. Healey and W.E. Blanz. Identifying metal surfaces in color images. In *SPIE Technical Symposium on Optics, Electro-Optics, and Sensors*, Orlando, Florida, April 4-8 1988. SPIE.

[63] G. Healey and R. Kondepudy. CCD camera calibration and noise estimation. In *IEEE Conference on Computer Vision and Pattern Recognition (CVPR'92)*, pages 90–95, Champaign, IL, June 1992. IEEE.

[64] G. Healey, S.A. Shafer, and L.B. Wolff. *Physics-based Vision: Principles and Practice - Color*. Jones and Bartlett, Boston, MA, 1992.

[65] J. Ho, B.V. Funt, and M.S. Drew. Separating a color signal into illumination and surface reflectance components: Theory and applications. *IEEE Trans. on Pattern Analysis and Machine Intelligence (PAMI)*, 12(10):966–977, Oct 1990.

[66] B.K.P. Horn. Understanding image intensities. *Artificial Intelligence*, 8(11):201–231, 1977.

[67] B.K.P. Horn. Exact reproduction of colored images. *Computer Vision, Graphics and Image Processing (CVGIP)*, 26:135–167, 1984.

[68] B.K.P. Horn. *Robot Vision*. MIT Electrical Engineering and Computer Science Series. MIT Press, Cambridge, Massachusetts London, England, 1986.

[69] B.K.P. Horn and M.J. Brooks. *Shape from Shading*. MIT Press, Cambridge, MA, 1989.

[70] B.K.P. Horn and B.G. Schunk. Determining optical flow. *Artificial Intelligence*, 17:185–203, 1981.

[71] S.L. Horowitz and T. Pavlidis. Picture segmentation by a tree traversal algorithm. *Journal of the ACM*, 23:368–388, 1976.

[72] A. Huertas and R. Nevatia. Detecting buildings in aerial images. *Computer Vision, Graphics and Image Processing (CVGIP)*, 41:131–152, 1988.

[73] R.S. Hunter. *The Measurement of Appearance*. John Wiley and Sons, New York, 1975. glossary.

[74] K. Ikeuchi. Determining surface orientations of specular surfaces by using the photometric stereo method. *IEEE Pattern Analysis and Machine Intelligence*, 3(6):661–669, November 1981.

[75] K. Ikeuchi. Shape from regular patterns. *Artificial Intelligence*, 22:49–75, 1984.

[76] K. Ikeuchi and B.K.P. Horn. Numerical shape from shading and occluding boundaries. *Artificial Intelligence*, 17:141–184, 1981.

[77] K. Ikeuchi and L. Sato. Determining reflectance properties of an object using range and brightness images. *IEEE Trans. on Pattern Analysis and Machine Intelligence (PAMI)*, 13(11):1115–1138, Nov 1991.

[78] R.B. Irvin and D.M. McKeown. Methods for exploiting the relationship between buildings and their shadows in aerial imagery. *IEEE Trans. on Systems, Man and Cybernetics*, 19(6):1564–1575, 1989.

[79] Y. Jang. Identification of interreflection in color images using a physics-based reflection model. In *IEEE Conference on Computer Vision and Pattern Recognition (CVPR'91)*, pages 632–637, Lahaina, Maui, Hawaii, June 1991. IEEE.

[80] F.A. Jenkins and H.E. White. *Fundamentals of Optics*. McGraw-Hill, New York, 1976.

[81] C. Jiang and M.O. Ward. Shadow identification. In *IEEE Conference on Computer Vision and Pattern Recognition (CVPR'92)*, pages 606–612, Champaign, IL, June 1992. IEEE.

[82] R.M. Johnston. Geometric metamerism. *Color Engineering*, 5(3):42, May-June 1967.

[83] R.M. Johnston. Color theory. In T.C. Patton, editor, *Pigment Handbook*, volume III, chapter III-D-b, pages 229–287. John Wiley and Sons, Inc., New York, 1974.

[84] R.M. Johnston and R.E. Park. Color and appearance. *Color Engineering*, 4(6):14, 1966.

[85] R. Johnston-Feller and D. Osmer. Exposure evaluation, part II – Bronzing. *Journal of the Coating Technology*, 51(650):37–44, March 1979.

[86] R.M. Johnston-Feller. Inhomogeneous materials vs. uniform materials. Private communication, 1987.

[87] D.B. Judd and G. Wyszecki. *Color in Business, Science and Industry*. John Wiley and Sons, New York, 1975.

[88] T. Kanade. Region segmentation: Signal vs. semantics. In *Proceedings of the 4th International Joint Conference in Pattern Recognition (IJCPR)*, pages 95–105, Kyoto, Japan, November 1978. IEEE.

[89] T. Kanade and K. Ikeuchi. Introduction to the special issue on physical modeling in computer vision. *IEEE Trans. on Pattern Analysis and Machine Intelligence (PAMI)*, 13(7):609–610, July 1991.

[90] T. Kanade and J.R. Kender. Mapping image properties into shape constraints: Skewed symmetry, affine-transformable patterns and the shape-from-texture paradigm. In A. Rosenfeld and J. Beck, editors, *Human and Machine Vision*. Academic Press, New York, 1983. Appeared also as technical report CMU-CS-80-133, Computer Science Department, Carnegie-Mellon University, Pittsburgh, PA.

[91] J.R. Kender. *Shape from Texture*. PhD thesis, Computer Science Department, Carnegie-Mellon University, November 1980. Appeared also as technical report CMU-CS-81-102, Computer Science Department, Carnegie-Mellon University, Pittsburgh, PA.

[92] J.R. Kender and E.M. Smith. Shape from darkness: Deriving surface information from dynamic shadows. In *Proceedings of the First International Conference on Computer Vision (ICCV)*, pages 539–546, London, June 1987. IEEE.

[93] G.J. Klinker. *A Physical Approach to Color Image Understanding*. PhD thesis, Computer Science Department, Carnegie-Mellon University, May 1988. Available as technical report CMU-CS-88-161.

[94] G.J. Klinker. We need interactive data interpretation rather than interactive data visualization. Workshop on Scientific Visualization, Visualization '91, San Diego, October 1991, 1991.

[95] G.J. Klinker. VDI – a Visual Debugging Interface for image interpretation and other applications. In F. H. Post and A.J.S. Hin, editors, *Advances in Scientific Visualization*. Springer Verlag, Berlin, Heidelberg, New York, 1992. Presented at the Second Eurographics Workshop on Visualization in Scientific Computing, Delft, Netherlands, 22-24 April, 1991. Also available as technical report CRL 91/2 from the Cambridge Research Lab, Cambridge, MA 02139.

[96] G.J. Klinker. An environment for empirical data interpretation. Submitted to Siggraph'93, August 1993.

[97] G.J. Klinker, S.A. Shafer, and T. Kanade. Measurement of gloss from color images. In *Inter-Society Color Council (ISCC) 87 Conference on Appearance*, pages 9–12, Williamsburg, VA, February 8-11 1987.

[98] G.J. Klinker, S.A. Shafer, and T. Kanade. Using a color reflection model to separate highlights from object color. In *Proceedings of the First International Conference on Computer Vision (ICCV)*, pages 145–150, London, June 1987. IEEE.

[99] G.J. Klinker, S.A. Shafer, and T. Kanade. The measurement of highlights in color images. *International Journal on Computer Vision (IJCV)*, 2(1):7–32, 1988.

[100] G.J. Klinker, S.A. Shafer, and T. Kanade. A physical approach to color image understanding. *International Journal on Computer Vision*, 4(1), 1990.

[101] R.D. Kriz. PV-Wave point and click. *PIXEL*, 2(2):28–30, 1991.

[102] E. Krotkov. Focusing. *International Journal on Computer Vision (IJCV)*, 1(3), 1987.

[103] J. Krumm and S.A. Shafer. Local spatial frequency analysis of image texture. In *Proceedings of the third International Conference on Computer Vision (ICCV)*, pages 354–358, Osaka, Japan, Dec 1990. IEEE.

[104] J. Krumm and S.A. Shafer. Shape from periodic texture using the spectrogram. In *IEEE Conference on Computer Vision and Pattern Recognition (CVPR'92)*, pages 284–289, Champaign, IL, June 1992. IEEE.

[105] M.S. Langer and S.W. Zucker. Qualitative shape from active shading. In *IEEE Conference on Computer Vision and Pattern Recognition (CVPR'92)*, pages 713–715, Champaign, IL, June 1992. IEEE.

[106] Y. LeClerc. A method for spectral linearization. Private communication, 1986.

[107] R.S. Ledley, M. Buas, and T.J. Golab. Fundamentals of true-color image processing. In *IEEE 10th International Conference on Pattern Recognition (ICPR'90)*, pages 791–795, Atlantic City, NJ, June 1990. IEEE.

[108] H.-C. Lee. Computing the scene illuminant color from specular highlight. Technical Report preprint, Eastman Kodak Research Laboratories, Eastman Kodak, Inc., Rochester, NY, 1986.

[109] H.-C. Lee. Method for computing the scene-illuminant chromaticity from specular highlights. *Journal of the Optical Society of America A (JOSA A)*, 3(10):1694–1699, October 1986.

[110] H.-C. Lee, E.J. Breneman, and C.P. Schulte. Modeling light reflection for computer color vision. *IEEE Trans. on Pattern Analysis and Machine Intelligence (PAMI)*, 12(4):402–409, Apr 1990.

[111] S.W. Lee. *Understanding of Surface Reflections in Computer Vision by Color and Multiple Views.* PhD thesis, GRASP Laboratory, Dep. of Computer and Information Science, University of Pennsylvania, Philadelphia, PA 19104, 1991.

[112] S.W. Lee. Highlight detection on cars. Private Communication, September 1992.

[113] S.W. Lee and R. Bajcsy. Detection of specularity using color and multiple views. In *Proceedings of the second European Conference on Computer Vision (ECCV)*, pages 99–114, Santa Margherita Ligure, Italy, May 1992. Springer Verlag.

[114] Y. Liow and T. Pavlidis. Use of shadows for extracting buildings in aerial images. *Computer Vision, Graphics and Image Processing (CVGIP)*, 49:242–277, 1990.

[115] D.G. Lowe and T.O. Binford. The interpretation of geometric structure from image boundary. In L.S. Bauman, editor, *DARPA-Image Understanding (IUS) workshop*, pages 39–41. Morgan Kaufmann, 1981.

[116] B. Lucas, G.D. Abram, D.A. Epstein, D.L Gresh, and K.P. McAuliffe. An architecture for a scientific visualization system. In *Proc. of Visualization '92*, pages 107–114, Boston, MA, October 1992. IEEE Computer Society Press.

[117] L.T. Maloney. Computational approaches to color constancy. Technical Report TR 1985-01, Stanford University, Applied Psychology Laboratory, January 1985.

[118] L.T. Maloney. Evaluation of linear models of surface spectral reflectance with small numbers of parameters. *Journal of the Optical Society of America A (JOSA A)*, 3(10):1673–1683, October 1986.

[119] L.T. Maloney and B.A. Wandell. Color constancy: A method for recovering surface spectral reflectance. *Journal of the Optical Society of America A (JOSA A)*, 3(1):29–33, January 1986.

[120] T.A. Mancini and L.B. Wolff. 3D shape and light source location from depth and reflectance. In *IEEE Conference on Computer Vision and Pattern Recognition (CVPR'92)*, pages 284–289, Champaign, IL, June 1992. IEEE.

[121] A.G. Maristany, P.K. Lebow, C.C. Brunner, D.A. Butler, and J.W. Funck. Classifying wood-surface features using dichromatic reflection. In *SPIE Conference on Optics in Agriculture and Forestry*, Boston, MA, November 1992. SPIE.

[122] D. Marr and E. Hildreth. Theory of edge detection. *Proc. R. Soc. Lond.*, B 207:187–217, 1980.

[123] C.C. McConnell and D.T. Lawton. IU software environments. In L.S. Bauman, editor, *DARPA-Image Understanding (IUS) workshop*, pages 666–677. Morgan Kaufmann, April 1988.

[124] P.J. Mercurio. The data visualizer. *PIXEL*, 2(2):31–35, 1991.

[125] P.J. Mercurio. Khoros. *PIXEL*, 3(1):28–33, 1992.

[126] P. Moon. A table of Fresnel reflection. *J. Math. Phys.*, 19(1), 1940.

[127] N. Mukawa. Estimation of shape, reflection coefficients and illuminant direction from image sequences. In *Proceedings of the third International Conference on Computer Vision (ICCV)*, pages 507–512, Osaka, Japan, Dec 1990. IEEE.

[128] J. Mundy, T. Binford, T. Boult, A. Hanson, R. Beveridge, R. Haralick, V. Ramesh, C. Kohl, D. Lawton, D. Morgan, K. Price, and T. Strat. The image understanding environment program. In *IEEE Conference on Computer Vision and Pattern Recognition (CVPR'92)*, pages 406–416, Champaign, IL, June 1992. IEEE.

[129] H.N. Nair and C.V. Stewart. Robust focus ranging. In *IEEE Conference on Computer Vision and Pattern Recognition (CVPR'92)*, pages 309–314, Champaign, IL, June 1992. IEEE.

[130] S.K. Nayar. Shape from focus. Technical Report CMU-RI-TR-89-27, Robotics Institute, Carnegie Mellon University, Pittsburgh, PA, 1989.

[131] S.K. Nayar. Shape from focus system. In *IEEE Conference on Computer Vision and Pattern Recognition (CVPR'92)*, pages 302–308, Champaign, IL, June 1992. IEEE.

[132] S.K. Nayar and Y. Gong. Colored interreflections and shape recovery. In L.S. Bauman, editor, *DARPA-Image Understanding (IUS) workshop*, San Diego, CA, Jan 1992. Morgan Kaufmann.

[133] S.K. Nayar, K. Ikeuchi, and T. Kanade. Determining shape and reflectance of hybrid surfaces by photometric sampling. *IEEE Trans. on Robotics Automation*, 6(4):418–431, Aug 1990.

[134] S.K. Nayar, K. Ikeuchi, and T. Kanade. Shape from interreflections. In *Proceedings of the third International Conference on Computer Vision (ICCV)*, pages 2–11, Osaka, Japan, Dec 1990. IEEE.

[135] S.K. Nayar, K. Ikeuchi, and T. Kanade. Surface reflection: Physical and geometrical perspectives. *IEEE Trans. on Pattern Analysis and Machine Intelligence (PAMI)*, 13(7):611–634, July 1991.

[136] S.K. Nayar, A.C. Sanderson, L.E. Weiss, and D.D. Simon. Specular surface inspection using structured highlights and Gaussian images. *IEEE Trans. on Robotics Automation*, 6(2):208–218, Apr 1990.

[137] S.K. Nayar, L.E. Weiss, D.A. Simon, and A.C. Sanderson. Structured highlight inspection of specular surfaces using extended Gaussian images. In *SPIE Conf. on Optics Illumination and Image Sensing for Machine Vision III*, volume 1005, Cambridge, MA, Nov 8-7 1988. SPIE.

[138] R. Nevatia. A color edge detector and its use in scene segmentation. *IEEE Transactions on Systems, Man and Cybernetics*, TSMC-7(11):820–826, Nov 1977.

[139] F.E. Nicodemus, J.C. Richmond, J.J. Hsia, I.W. Ginsberg, and T. Limperis. Geometrical considerations and nomenclature for reflectance. Technical Report NBS Monograph 160, National Bureau of Standards, Washington, DC, October 1977.

[140] C.L. Novak. *Estimating Scene Properties by Analyzing Color Histograms with Physics-Based Models*. PhD thesis, School of Computer Science, Carnegie Mellon, December 1992.

[141] C.L. Novak and S.A. Shafer. Supervised color constancy using a color chart. Technical Report CMU-CS-TR-90-140, School of Computer Science, Carnegie Mellon, Pittsburgh, PA, 1990.

[142] C.L. Novak and S.A. Shafer. Anatomy of a histogram. Technical Report CMU-CS-TR-91-203, School of Computer Science, Carnegie Mellon, Pittsburgh, PA, Nov 1991.

[143] C.L. Novak and S.A. Shafer. Anatomy of a histogram. In *IEEE Conference on Computer Vision and Pattern Recognition (CVPR'92)*, pages 599–605, Champaign, IL, June 1992. IEEE.

[144] C.L. Novak and S.A. Shafer. Estimating scene properties from color histograms. Technical Report CMU-CS-92-212, School of Computer Science, Carnegie Mellon, Pittsburgh, PA, Nov 1992.

[145] C.L. Novak, S.A. Shafer, and R.G. Willson. Obtaining accurate color images for machine vision research. In *SPIE Conf. on Perceiving, Measuring and Using Color*, volume 1250. SPIE, Feb 1990.

[146] R. Ohlander, K. Price, and D.R. Reddy. Picture segmentation using a recursive region splitting method. *Computer Graphics and Image Processing*, 8:313–333, 1978.

[147] Y. Ohta, T. Kanade, and T. Sakai. Color information for region segmentation. *Computer Graphics and Image Processing*, 13:222–231, 1980.

[148] CIE (International Commission on Illumination). International lighting vocabulary, 3rd edition. Technical Report 17 (E-1.1), CIE, Bureau Central de la CIE, 4 Av. du Recteur Poincaré, 75-Paris 16, France, 1970.

[149] J.-S. Park and J.T. Tou. Highlight separation and surface orientations for 3-D specular objects. In *IEEE 10th International Conference on Pattern Recognition (ICPR'90)*, pages 331–335, Atlantic City, NJ, June 1990. IEEE.

[150] T. Pavlidis. *Structural Pattern Recognition*. Springer Verlag, Berlin, Heidelberg, New York, 1977.

[151] A. Pentland. Photometric motion. *IEEE Trans. on Pattern Analysis and Machine Intelligence (PAMI)*, 13(9):879–890, Sep 1991.

[152] B.T. Phong. Illumination for computer generated pictures. *Communications of the ACM*, 18:311–317, 1975.

[153] L. Quam. The Image Calc vision system. Technical report, Stanford Research Institute, Menlo Park, CA, 1984.

[154] K. Riley and C. McConnell. Powervision. Technical report, Advanced Decision Systems, Mountain View, CA, March 1988.

[155] L.G. Roberts. Machine perception of three-dimensional solids. In Tippet et al., editor, *Optical and Electro-Optical Image Processing*, chapter 9, pages 159–197. MIT Press, Cambridge, Mass., 1965.

[156] A. Rosenfeld and A.C. Kak. *Digital Picture Processing*, volume 2. Academic Press, New York, 1982. 2. edition.

[157] J.M. Rubin and W.A. Richards. Color vision and image intensities: When are changes material? *Biological Cybernetics*, 45:215–226, 1982.

[158] A.C. Sanderson, L.E. Weiss, and S.K. Nayar. Structured highlight inspection of specular surfaces. *IEEE Trans. on Pattern Analysis and Machine Intelligence (PAMI)*, 10(1):44–55, Jan 1988.

[159] C. Schroeder. Ein Ansatz zur Segmentation von Farbbildern durch hierarchische Ballungsanalyse. Technical Report FBI-HH-B-132/87, Universitaet Hamburg, Fachbereich Informatik, Bodenstedtstrasse 16, D-2000 Hamburg 50, October 1987.

[160] S.A. Shafer. Describing light mixtures through linear algebra. *Journal of the Optical Society of America (JOSA A)*, 72(2):299–300, February 1982.

[161] S.A. Shafer. Optical phenomena in computer vision. In *CSCSI-84*, Ontario, May 1984. Canadian Society for Computational Studies of Intelligence. Also available as technical report TR-135, Computer Science Department, University of Rochester, March 1984.

[162] S.A. Shafer. The Calibrated Imaging Lab under construction at CMU. In L.S. Bauman, editor, *Proceedings of the DARPA Image Understanding Workshop, held in Miami Beach*, pages 509–515, November 1985.

[163] S.A. Shafer. Using color to separate reflection components. *COLOR research and application*, 10(4):210–218, Winter 1985. Also available as technical report TR-136, Computer Science Department, University of Rochester, NY, April 1984.

[164] S.A. Shafer and R.M. Johnston-Feller. Current problems with gloss traps and how to avoid them. Private communication, 1987.

[165] S.A. Shafer and T. Kanade. Recursive region segmentation by analysis of histograms. In *Proceedings of the International Conference on Acoustics, Speech and Signal Processing*, pages 1166–1171, Paris, France, May 1982. IEEE.

[166] S.A. Shafer and T. Kanade. Using shadows in finding surface orientations. *Computer Vision, Graphics and Image Processing (CVGIP)*, 22:182–199, 1983.

[167] S.A. Shafer, T. Kanade, G.J. Klinker, and C.L. Novak. Physics-based models for early vision by machine. In *SPIE Conf. on Perceiving, Measuring and Using Color*, volume 1250. SPIE, Feb 1990.

[168] M. Shao, T. Simchony, and R. Chellappa. New algorithms for reconstruction of a 3-D depth map from one or more images. In *IEEE Conference on Computer Vision and Pattern Recognition (CVPR'88)*, pages 530–535, Ann Arbor, MI, June 1988. IEEE.

[169] J.F. Snell. Radiometry and photometry. In W.G. Driscoll and W. Vaughan, editors, *Handbook of Optics*. McGraw-Hill, New York, 1978.

[170] F. Solomon and K. Ikeuchi. Extracting the shape and roughness of specular lobe objects using four-light photometric stereo. In *IEEE Conference on Computer Vision and Pattern Recognition (CVPR'92)*, pages 466–471, Champaign, IL, June 1992. IEEE.

[171] W.N. Sproson. *Colour Science in Television and Display Systems*. Adam Hilger Ltd, Bristol, 1983.

[172] T. Stephenson. Computer graphics and computer vision. *Advanced Imaging*, May 1990.

[173] M. Subbarao and T.-C. Wei. Depth from defocus and rapid autofocusing: a practical approach. In *IEEE Conference on Computer Vision and Pattern Recognition (CVPR'92)*, pages 479–484, Champaign, IL, June 1992. IEEE.

[174] M.J. Swain. Color indexing. Technical Report TR 360, Computer Science Department, University of Rochester, Rochester, NY, Nov 1990.

[175] M.J. Swain and D.H. Ballard. Indexing via color histograms. In *Proceedings of the third International Conference on Computer Vision (ICCV)*, pages 390–393, Osaka, Japan, Dec 1990. IEEE.

[176] M.J. Swain and D.H. Ballard. Color indexing. *International Journal on Computer Vision (IJCV)*, 7(1):11–32, Nov 1991.

[177] H.D. Tagare and R.J.P. deFigueiredo. A framework for the construction of general reflectance maps for machine vision. Technical Report EE88-16, Dept. of Elect. Comp. Eng., Rice University, Apr 1988.

[178] H.D. Tagare and R.J.P. deFigueiredo. A theory of photometric stereo for a general class of reflectance maps. In *IEEE Conference on Computer Vision and Pattern Recognition (CVPR'89)*, pages 38–45, San Diego, CA, June 1989. IEEE.

[179] H.D. Tagare and R.J.P. deFigueiredo. Simultaneous estimation of shape and reflectance maps from photometric stereo. In *Proceedings of the third International Conference on Computer Vision (ICCV)*, pages 340–343, Osaka, Japan, Dec 1990. IEEE.

[180] H.D. Tagare and R.J.P. deFigueiredo. A theory of photometric stereo for a class of diffuse non-Lambertian surfaces. *IEEE Trans. on Pattern Analysis and Machine Intelligence (PAMI)*, 13(2):133–152, Feb 1991.

[181] R.S. Thau. Illuminant precompensation for texture discrimination using filters. In L.S. Bauman, editor, *DARPA-Image Understanding (IUS) workshop*, pages 174–178. Morgan Kaufmann, Sep 1990.

[182] C.E. Thorpe. *FIDO: Vision and Navigation for a Robot Rover*. PhD thesis, Computer Science Department, Carnegie-Mellon University, December 1984. Available as technical report CMU-CS-84-168.

[183] S. Tominaga. A color classification method for color images using a uniform color space. In *IEEE 10th International Conference on Pattern Recognition (ICPR'90)*, pages 803–807, Atlantic City, NJ, June 1990. IEEE.

[184] S. Tominaga. Surface identification using the dichromatic reflection model. *IEEE Trans. on Pattern Analysis and Machine Intelligence (PAMI)*, 13(7):658–670, July 1991.

[185] S. Tominaga and B.A. Wandell. The standard surface reflectance model and illuminant estimation. *Journal of the Optical Society of America A (JOSA A)*, 6(4):576–584, 1989.

[186] S. Tominaga and B.A. Wandell. Component estimation of surface spectral reflectance. *Journal of the Optical Society of America A (JOSA A)*, 7(2):312–317, 1990.

[187] F. Tong and B.V. Funt. Specularity removal for shape from shading. In *Proc. of the Conference Vision Interface, Edmonton, Alberta, Canada*, 1988.

[188] K.E. Torrance and E.M. Sparrow. Theory for off-specular reflection from roughened surfaces. *Journal of the Optical Society of America A (JOSA A)*, 57:1105–1114, September 1967.

[189] M. Tsukada and Y. Ohta. An approach to color constancy using multiple images. In *Proceedings of the third International Conference on Computer Vision (ICCV)*, pages 385–389, Osaka, Japan, Dec 1990. IEEE.

[190] S. Ullman. The interpretation of structure from motion. AI Memo 476, MIT AI Laboratory, Cambridge, Mass., October 1976.

[191] C. Upson, T. Faulhaber Jr., D. Kamins, D. Laidlaw, D. Schlegel, J. Vroom, R. Gurwitz, and A. van Dam. The Application Visualization System: A computational environment for scientific visualization. *IEEE Computer Graphics and Applications*, 9(4):30–42, 1989.

[192] M. Vriesenga, G. Healey, K. Peleg, and J. Sklansky. Controlling illumination color to enhance object discriminability. In *IEEE Conference on Computer Vision and Pattern Recognition (CVPR'92)*, pages 710–712, Champaign, IL, June 1992. IEEE.

[193] R.S. Wallace. Minimal entropy clustering. Thesis Proposal, Computer Science Department, Carnegie Mellon University, 1987.

[194] T. Westman, D. Harwood, T. Laitinen, and M. Pietikaeinen. Color segmentation by hierarchical connected components analysis with image enhancement by symmetric neighborhood filters. In *IEEE 10th International Conference on Pattern Recognition (ICPR'90)*, pages 796–802, Atlantic City, NJ, June 1990. IEEE.

[195] T.D. Williams. Image understanding tools. In *IEEE 10th International Conference on Pattern Recognition (ICPR'90)*, pages 606–610, Atlantic City, NJ, June 1990. IEEE.

[196] S.J. Williamson and H.Z. Cummins. *Light and Color in Nature and Art*. John Wiley and Sons, New York, 1983.

[197] L.B. Wolff. Spectral and polarization stereo methods using a single light source. In *Proceedings of the First International Conference on Computer Vision (ICCV)*, pages 708–715, London, June 1987. IEEE.

[198] L.B. Wolff. Classification of material surfaces from the polarization of specular highlights. In *SPIE Conf. on Optics, Illumination and Image Sensing For Machine Vision III*, pages 206–213. SPIE, 1988.

[199] L.B. Wolff. Material classification and separation of reflection components using polarization/radiometric information. In L.S. Bauman, editor, *DARPA-Image Understanding (IUS) workshop*. Morgan Kaufmann, May 1989.

[200] L.B. Wolff. Shape understanding from Lambertian photometric flow fields. In *IEEE Conference on Computer Vision and Pattern Recognition (CVPR'89)*, pages 46–52, San Diego, CA, June 1989. IEEE.

[201] L.B. Wolff. Using polarization to separate reflection components. In *IEEE Conference on Computer Vision and Pattern Recognition (CVPR'89)*, pages 363–369, San Diego, CA, June 1989. IEEE.

[202] L.B. Wolff. Polarization-based material classification from specular reflection. *IEEE Trans. on Pattern Analysis and Machine Intelligence (PAMI)*, 12(11):1059–1071, Nov 1990.

[203] L.B. Wolff. *Polarization methods in computer vision*. PhD thesis, Columbia University, Jan 1991.

[204] L.B. Wolff. Diffuse reflection. In *IEEE Conference on Computer Vision and Pattern Recognition (CVPR'92)*, pages 472–478, Champaign, IL, June 1992. IEEE.

[205] L.B Wolff and T. Boult. Polarization/radiometric methods for material classification. In *IEEE Conference on Computer Vision and Pattern Recognition (CVPR'89)*, pages 387–395, San Diego, CA, June 1989. IEEE.

[206] L.B Wolff and T. Boult. Constraining object features using a polarization reflectance model. *IEEE Trans. on Pattern Analysis and Machine Intelligence (PAMI)*, 13(7):635–657, July 1991.

[207] L.B. Wolff, S.A. Shafer, and G. Healey. *Physics-based Vision: Principles and Practice - Radiometry*. Jones and Bartlett, Boston, MA, 1992.

[208] L.B. Wolff, S.A. Shafer, and G. Healey. *Physics-based Vision: Principles and Practice - Shape Recovery*. Jones and Bartlett, Boston, MA, 1992.

[209] R.J. Woodham. *Reflectance Map Techniques for Analyzing Surface Defects in Metal Castings*. PhD thesis, Artificial Intelligence Laboratory, Massachusette Institute of Technology, June 1978.

[210] R.J. Woodham. Multiple light source optical flow. In *Proceedings of the third International Conference on Computer Vision (ICCV)*, pages 42–46, Osaka, Japan, Dec 1990. IEEE.

[211] Y. Yang and A. Yuille. Sources from shading. In *IEEE Conference on Computer Vision and Pattern Recognition (CVPR'91)*, pages 534–539, Lahaina, Maui, Hawaii, June 1991. IEEE.

[212] S. Yi, R.M. Haralick, and L.G. Shapiro. Automatic sensor and light source positioning for machine vision. In *IEEE 10th International Conference on Pattern Recognition (ICPR'90)*, pages 55–59, Atlantic City, NJ, June 1990. IEEE.

[213] Q. Zheng and R. Chellappa. Estimation of illuminant direction, albedo and shape from shading. *IEEE Trans. on Pattern Analysis and Machine Intelligence (PAMI)*, 13(7):680–702, July 1991.

Index

active vision, 125

bidirectional spectral-reflectance distribution function (BSRDF), 120

color cluster analysis
  from color variation on an entire object ("global method"), 16–22, 52, 69, 72–82, 86, 89–92, 95, 97
  from local color variation ("local method"), 5, 22–29, 52–68, 72, 77–82, 85–86, 89–92, 97–99, 118
color cluster shape, *see* histogram shape
color constancy, 3, 7, 46, 92, 108, 111, 122–124
color cube, 37–38, 46–47, 57, 61, 64, 72–73, 75, 83, 93, 102, *see also* color histogram
  dark corner, 61, 95

color histogram, 38, 41, 43, 46, 52, 69, 108, 119, 121, *see also* color space
color indexing, 119
color mixture, 5
color restoration for clipped and bloomed pixels, 35, 84–87
color space, 3, 5, 36–37, 43, 45, 52, 56, 59–60, 63, 65, 69, 72–73, 81, 93–94, 102, 114–115, 118, 121, *see also* vector space
  CIELAB, 121
  IHS, 6, 93, 114, 118, 121
  origin, 73, 93, 102
  RGB, 6, 114, 121
computer graphics, 11–12, 122, 126
computer vision, 1–3, 9, 12, 15, 71, 105, 107, 112–113
  physics-based, 3, 7, 113, 122, 125–128
  traditional, 2–4, 125–126

continuous light spectrum, 36, 121

edge detection, 2, 6, 118

Fresnel ratio, 117
future work (1988), 13, 40, 44, 69,
    76, 84, 100, 109–112

Gaussian spectrum sphere, 23–27
    difference vectors as points on
        great circle, 25
    great circle, 25
    highlight clusters as partial
        great circles, 27
    inside segment of great cir-
        cle, 25
    matte clusters as points, 24
    outside segment of great cir-
        cle, 25, 27
geometric optics, 31, 115

heuristics, 4, 7, 68, 74–75, 82, 94,
    98–100, 117
    50%-heuristic, 18, 22, 63, 100,
        110, 129–132
    blooming heuristic, 79–80, 94
    dark heuristic, 61–62, 96
    highlight heuristic, 99
    Lambertian consistancy, 114
    matte heuristic, 98–99
    proximity heuristic, 62, 67,
        69, 96
    restoration heuristic, 85
    too bright intensity, 116
    too sharp shading peak, 117
hierarchical connected component
    analysis, 118
highlight detection, 84, 86, 92, 108,
    115–117, 126, see also pixel
    splitting

highlight removal, 84, 86–87, 92,
    108, 117, 126, see also
    pixel splitting
highlight separation, see pixel split-
    ting
histogram backprojection, 119
histogram intersection, 119
histogram shape, 5, 105
    bloomed color pixel, 40, 46,
        65, 72, 80–81, 84–87, 94,
        98
    clipped cluster, 38–39, 46, 65,
        74, 85
    cluster dimensionality, 56–58,
        93
    collinear matte and highlight
        clusters, 46, 81–82, 101
    collinear matte color clusters,
        101, 111
    color cylinder, 60–61, 64, 93,
        96–97, 101
    color cylinder intersection for
        objects with similar col-
        ors, 62
    concavity between matte and
        highlight clusters, 74
    coplanar clipped cluster and
        dichromatic plane, 40
    coplanar color clusters, 66–
        70
    coplanar matte clusters, 57
    dark sphere, 61
    dichromatic plane, 5, 16–18,
        25, 27, 29, 37–40, 57, 59,
        63, 65–69, 72–73, 75, 83–
        85, 93, 99, 102
    diffuse lobe, 120
    dog leg, 18
    foothill of highlight cluster,
        81–82
    highlight cluster, 18, 39, 46–
        47, 56, 59–60, 62–65, 68,

73–74, 80–81, 99, 101–
    102, 111, 114–115
highlight line, 17, 64, 74–75,
    77, 79, 85
interreflection cluster, 80, 82,
    121
length of highlight cluster, 81–
    82
length of linear cluster, 81
length of matte cluster, 81–
    82
length of matte line, 20, 64
linear cluster, 16, 32, 46, 56–
    60, 62, 73, 82, 96, 98
matte cluster, 43, 46–47, 53,
    57, 59–65, 67–70, 73–74,
    79–81, 95–96, 98–99, 101–
    102, 111, 115
matte cylinder, 64, 67, 69,
    81, 96–97, 101
matte line, 17–18, 22, 24, 64,
    74–77, 79, 85, 100
missing highlight cluster, 102
missing matte cluster, 102
non-linear cluster, 80
orientation of clipped cluster,
    38, 40
orientation of highlight clus-
    ter, 40, 46, 65, 75
orientation of highlight line,
    25, 74–76, 81
orientation of linear cluster,
    60, 73
orientation of matte cluster,
    40, 75, 81
orientation of matte line, 24,
    75
parallel highlight clusters, 27,
    46, 65, 90
parallelogram, 16, 37, 100
planar cluster, 57–60, 63
planar slice, 66, 68, 96

pointlike cluster, 56, 58–60
position of linear cluster, 60,
    73
several highlight clusters, 100
single linear cluster, 102, 111
skewed L, 18, 130
skewed T, 5, 18, 41, 52–54,
    60, 63–65, 71–73, 77, 99,
    121
skewed wedge, 18
skewing angle of T, 18
sparse highlight cluster, 79–
    81
specular lobe, 115, 120–121
specular spike, 115, 120–121
start of highlight cluster, 74,
    80–81
start of highlight line, 18, 20,
    22, 65, 81, 85, 100, 111
volumetric cluster, 58, 60
hypotheses, 5–7, 53, 55, 72, 77,
    106–107, 111, 118
camera problem hypothesis,
    111
candidate matte area, 60
competing hypotheses, 96–97
dielectric hypothesis, 103
false linear hypothesis, 98
highlight candidate, 63–65, 99
highlight hypothesis, 65
illumination hypothesis, 65
initial hypothesis, 59–60
linear hypothesis, 55, 60–63,
    65, 93, 97–99
matte hypothesis, 98
metallic hypothesis, 103
planar hypothesis, 55, 59, 63,
    65, 67–68, 93, 98
previous hypothesis, 97
refinement, 96–97
similar hypotheses, 97

illuminants
  blue sky, 123, 126
  grey light, 14
  narrow-band illuminant, 102
  pink light, 89–90, 92, 98
  tungsten, 41, 77
  white light, 14, 47, 52, 66,
    69, 77, 83, 89–90, 92–93,
    96, 106
  yellow light, 46, 89–90, 92–
    93, 96, 102
illumination environment, 1, 13,
    45, 89, 106, 108–109, 115,
    123, 126
  ambient light, 13, 46, 93, 106,
    109, 123, 126
  body-to-body interreflection,
    122
  body-to-surface interreflection,
    122
  diffuse interreflection, 122
  diffuse-to-specular interreflec-
    tion, 122
  gloss trap, 126
  hair dryer, 115, 128
  illumination color, 6–7, 45–
    46, 54, 65, 77, 90, 92–94,
    102, 106, 111, 114, 119–
    120, 122
  illumination direction, 13, 16–
    19, 57, 115, 125, 129
  illumination position, 16, 18–
    19, 32, 99, 107–108, 111,
    129, 139
  interreflection, 13, 68–69, 79–
    80, 82, 85, 96–97, 109,
    111, 114–115, 118, 121–
    123, 126–127
  interreflection with multiple
    bounces, 122
  light source position, 130
  light-diffusing sphere, 115

  photometric stereo, 115, 124,
    128
  secondary illumination, 122–
    123
  several light sources, 109, 115,
    125–126
  shadows, 1–3, 27, 51, 68–69,
    109, 111, 114, 123, 126
  single light source, 13, 19, 30,
    32, 115
  spectrophotometer, 126
  spectroradiometer, 77, 90, 120
  specular interreflection, 122
  specular-to-specular interreflec-
    tion, 122
  surface-to-body interreflection,
    122
  surface-to-surface interreflec-
    tion, 122
image areas, 59–70
  area containing edges, 118
  area from rough surfaces, 102
  area inconsistent with Dichro-
    matic Reflection Model,
    5, 84, 109, 111
  area with small amount of gloss,
    86
  area with unreliable color in-
    formation, 5, 38, 40, 45,
    60, 86, 95–96, 101, 111
  bloomed area, 111
  bright matte area, 6–7, 93,
    97, 106
  dark area, 94–95, 101
  dark matte area, 6–7, 56, 93,
    97, 106
  edges, 118
  entire object area, 46, 58, 94,
    96
  from physics-based segmen-
    tation, 59, 61–63, 66, 68–
    69, 71–73, 77, 82–84

from traditional segmentation, 3, 6, 54, 93

highlight area, 2–3, 5–7, 10, 16–18, 22, 24, 27, 47, 53–54, 56, 58–59, 62–63, 65, 67, 81, 86–87, 92–93, 98–100, 106–108, 110–111, 116–117

highlight boundary, 2, 46, 51, 89, 106, 108, 118

highlight center, 46, 58, 60, 65, 68, 85–86

homogeneous area, 6

inhomogeneous area, 6

interreflection area, 97, 111, 122, 127

Lambertian area, 116

large highlight area on flat object, 76, 98

local area, 53

material boundary, 2, 5, 18, 23, 27, 51–52, 58–60, 62–64, 85, 87, 89, 96, 98, 101, 106–108, 118

matte area, 5–6, 10, 18, 43, 53, 56, 58–59, 64, 66–67, 86, 93, 98–99, 106, 108, 116

mirror image, 2, 69

neighboring area, 57, 59, 63

neighboring highlight area, 63

neighboring matte area, 63, 65

neighboring object area, 97

non-Lambertian area, 116

object boundary, 2, 7, 87

occluding boundary, 57

partially shadowed area, 123

planar area, 68

shading boundary, 2, 5, 51, 111, 118

shadow area, 3, 111, 123

shadow boundary, 2, 51

transition area between matte pixels and bright highlight, 86, 108

image enhancement, 118

image segmentation
final, 68, 93
linear, 63, 65, 67, 69, 93
linear region growing, 61, 67
Phoenix, 6–7, 93, 118, 126
physics-based, 5–7, 51–71, 93–94, 105–107, 126
planar, 68–69, 93
planar region growing, 66–67, 70
region expansion, 68
region growing, 52–53
region merging, 52
region shrinking, 68
region splitting, 6–7, 52, 111, 118
traditional, 2, 6–7, 51, 54, 93–94, 105–107, 126

industrial applications, 110, 118, 125–128

intrinsic components, 5, 105
body reflection component, 14, 16–22, 24–25, 27–31, 46, 53, 56, 60, 69, 71–73, 79, 82–87, 99, 102, 108, 110, 115, 117, 122, 129–132, 139

diffuse component, 115, 120

highlight component, 7, 27, 53, 93, 108, 110

Lambertian component, 115–117, 120

matte component, 7

shading component, 53, 56, 81, 85–86, 110, 124

specular component, 115, 117

specular lobe component, 115,
    120
specular spike component, 115,
    120
surface reflection component,
    14–18, 24–25, 28–30, 33,
    43, 46, 53, 56, 60–61, 71–
    72, 77, 82–87, 92, 99, 102,
    107–108, 115, 117, 120,
    122, 130
intrinsic images, 5, 7, 53–54, 92,
    107, 114–117
    body reflection image, 6–7,
        71, 82–87, 92, 102, 107–
        108, 110, 114–117
    error image, 84, 86, 111
    highlight image, 107, see also
        surface reflection image
    noise image, 83–85, 111
    surface reflection image, 6–7,
        71, 82–87, 92, 107, 110,
        114–117
intrinsic models, see reflection mod-
    els

line splitting, 73–76, 95
linear combination of color vec-
    tors, 5, 14, 25, 32, 36,
    85

material characteristics, 1
    absorption power, 31
    albedo, 115, 125
    bronzing, 33
    conducting material, 30
    crystal, 31
    dielectric material, 30–31, 111,
        117, 120, 123–124, 127
    distribution of scattering par-
        ticles, 31
    dyed dielectrics, 10, 12, 31
    electric conductivity, 30

extinction coefficient, 30
fibers, 10, 12, 31
fluorescent material, 32
glossiness, 115
glossy material, 1, 102
inhomogeneous material, 31
interface, 11, 120
interference patterns, 33
material body, 9, 12–13, 30–
    31, 33, 110, 118
material layers, 33
material medium, 10–13, 33
material surface, 9, 11–13, 16,
    30–31, 33, 118
material thickness, 31
matte material, 1, 102
metal, 13, 30, 102, 110–111,
    117, 120–121, 124
metal-like dielectrics, 33
metallic flakes, 33
non-dielectric material, 30–31
non-uniform dielectrics, 31–
    32
non-uniform opaque dielectrics,
    13–15, 29–30, 32, 45–47,
    65, 77, 89, 92, 102–103,
    110
non-uniform transmittant di-
    electrics, 32
opacity, 10, 31–32, 41, 126
pigmented dielectrics, 10, 33
pigments, 10–13, 31, 33, 96,
    110
reflectance, 30, 42, 122–123
reflectance power, 30
self-shadowing, 11
surface roughness, 1, 11, 13,
    16, 31, 102, 115, 120–
    122, 126
uniform dielectrics, 31–32
uniformity scale, 31

material classification, 102–103, 108, 111, 117

measurement space, 3–5, 29, 115, 120–121, *see also* vector space

metamerism, 37

normalized color pixels, 25, 118

object characteristics
  concave object, 99–100, 110
  cylindrical object, 16–18, 23, 100, 110
  dark object, 52, 56, 72, 96–97, 102
  edges, 85, 89, 98, *see also* image areas: material boundary, image areas: occluding boundary, image areas: shading boundary
  ellipsoidal object, 99–100, 110
  flat object, 52, 56, 69, 72, 76, 97–98, 100, 110
  material, *see* material characteristics
  mirror, 126, *see also* image areas: mirror image
  object color, 5–7, 53, 66, 85–86, 97, 101–102, 106, 108, 111, 119, 123, 125, *see also* color constancy, intrinsic components, optical processes
  saturated colors, 81
  shape, 1, 96, 105
  spherical object, 18–29, 100, 110, 129–132
  surface curvature, 56
  surface roughness, *see* material characteristics: surface roughness

textured object, 3, 110, 124, 126, *see also* image areas: material boundary

optical processes, 2, 9–10, 12–13, 31, 89, 115
  absorption, 9–12, 30–31
  diffusion, 13, 16
  reflection, 6, 10, 12–13, 30, 35, 37, 51, 53–54, 116, 124
  refraction, 9–12, 31, 43, 46
  scattering, 11
  transmission, 12, 31–32, 36

parameters
  control parameters, 40–41, 56, 60–62, 64–66, 69, 73, 84, 94–98, 107
  hyperspace of segmentation parameters, 118
  physics-based scene parameters, 126
  wealth of physics-based scene parameters, 118, 126
photometric sampling, 115, 117
physical optics, 31, 115
pixel classes, 60, 73
  bloomed color pixel, 40, 46, 65, 72, 80–81, 84–87, 94, 98, 127
  bloomed highlight pixel, 85
  bloomed matte pixel, 85
  clipped color pixel, 38, 40, 46, 65, 72, 74–75, 77, 84–87, 94, 98, 111, 127
  false matte pixel, 75, 81
  highlight pixel, 16–18, 24–25, 27, 29, 57–58, 60–61, 64, 68, 73, 75, 77, 79, 81, 86, 94, 116, 127
  interreflection pixel, 116
  material boundary pixel, 87

matte pixel, 16–18, 24–25, 27–
    29, 43, 57–58, 60, 63–64,
    67, 73, 75, 77, 79, 94, 97,
    99, 127
non-Lambertian pixel, 116
specular pixel, 116
too dark pixel, 127
pixel splitting, 5–7, 53–54, 71–87,
    101–102, 105, 107–108,
    110, 114–117
polarization, 11, 116–117, 121, 124,
    128
principal component analysis, 56,
    72, 75, 80–81

radiosity, 122
range data, 115, 125
ray tracing, 122
real materials/media
    aluminum, 30
    brass, 120
    carpet, 126
    ceramics, 10, 31, 45, 47, 120
    copper, 120
    dye, 31
    fog, 126
    fruit, 120
    gemstone, 31
    glass, 31, 43
    gold, 30
    grass, 126
    leaves, 120
    orange peel, 120
    paint, 10, 31, 110, 120, 125
    paper, 10, 31, 45, 47, 120
    plastic, 7, 10, 31, 45–47, 52,
        89, 120
    plating, 120
    quartz, 31
    silver, 30
    smoke, 126
    textiles, 10, 12, 31, 120

tiles, 120
vinyl, 120
water, 31
wood, 120, 126–127
real scenes
    kitchen scene, 110, 125
    natural terrain, 118
    office scene, 110
    outdoor scene, 109, 118, 125
reflectance map, 107
reflection component, see intrin-
    sic component
reflection image, see intrinsic im-
    age
reflection models, 2–5, 7, 105–107,
    124
    Dichromatic Reflection Model
        (DRM), 11–16, 23, 27–
        30, 32, 35–38, 41–43, 46–
        47, 51–53, 60, 63, 68, 71–
        72, 84–85, 89, 94, 119–
        120, 122
    geometric optics, 115
    Neutral Interface Reflection
        Model (NIR), 120
    physical optics, 115
reflection process, see optical pro-
    cesses
reflection terms
    body reflection, 10, 12–13, 16–
        19, 30–32, 53, 60, 64, 71,
        108, 111, 114, 120, 122,
        130–131
    diffuse reflection, 12–13, 115–
        116, 124
    Fresnel reflection, 11
    glossy reflection, 12
    highlight reflection, 1–5, 7, 23,
        27–28, 32, 52–53, 89, 102,
        105–108, 115–116, 118,
        121–122, 126–127
    interface reflection, 11

interreflection, *see* illumination environment
Lambertian reflection, 16, 19,
114, 116
mirror reflection, 11, 13, 16,
99, 116
shading, 1–3, 5, 7, 23, 27–28,
32, 52–53, 89, 92–93, 96,
105–108, 110, 114, 117–
118, 121–122
specular reflection, 11–13, 116,
120, 124–125, 127
surface reflection, 10–14, 16–
17, 30, 32, 46, 53, 60,
63–65, 69, 71, 79, 100,
108, 111, 114, 117, 120,
122
reflection vectors
body reflection vector, 14, 16–
18, 24, 32, 36, 38–40, 43,
47, 66, 69, 71–87, 90, 95–
97, 99–102
color difference between body
and surface reflection, 101–
102
surface reflection vector, 15–
16, 18, 24, 27, 32, 36,
38–40, 47, 71–87, 90, 92,
99–102

sensor characteristics, 4, 52–53,
68, 93–94, 108–109, 111,
121
blooming, 5, 40, 46–47, 53,
58, 65, 68, 81, 84–85
chromatic aberration, 5, 43–
45, 47, 85, 121
color clipping, 38–40, 46, 54,
58, 65, 68, 81, 84–85, 94–
95
color dispersion, 43–44

dynamic range, 5, 37–41, 46,
94, 111
filters, 35–36, 41, 116
focal length, 43–44, 121, 127
gamma-correction, 5, 42–43
magnification, 44–45, 47
noise, 2, 4, 6, 41, 47, 51–53,
56, 58, 60, 66, 73, 79,
81–85, 87, 94, 101–102,
106, 111, 117
overloaded sensor element, 40
sensor orientation, 57
sensor position, 4, 16, 19, 107,
129–130
spatial averaging, 40
spectral integration, 36–37
spectral responsivity of sensor, 36, 41–43
technology, 5, 35, 37, 40–43,
56, 77, 90, 94, 108, 116,
120, 126
sensor fusion, 4
sensor modeling, 4–5, 51, 53, 105,
108–109
aperture control, 41–42, 77,
84, 111
color balancing, 41–42
magnification control, 44
spectral linearization, 41–43,
43–103
shape from X, 3–5, 84, 124
body reflection, 86, 92, 110
darkness, 123
focus, 127
gloss, 108
highlights, 7, 71, 86, 92, 103,
107, 110–111
interreflections, 127
Lambertian and specular reflection, 117
motion, 3, 7, 71, 86, 107, 124
motion and shading, 125

optical flow, 3, 124
photometric stereo, 113, 115,
    124, 128
photometric stereo with high-
    lights and albedo changes,
    124
shading, 7, 71, 84, 86, 92,
    103, 107, 110–111, 113,
    117, 122, 124–125, 127
shading and texture, 125
shadows, 123
stereo, 3, 7, 71, 86, 107, 124
stereo and highlights, 117, 125
stereo and shading, 125
surface reflection, 86, 92, 110
spectral crosspoint criterion, 23,
    27–29
spectral differencing, 114
spectral histogram, 16, 131, *see
    also* color histogram
specular sharpness, 115
specular stereo, 117
specular strength, 115
supervised color constancy, 123,
    128
symmetric neighborhood filter, 118

three-dimensional data, 115

vector space
    dimensionality, 4–5, 14, 16,
        23, 27, 29, 32, 36–37, 56–
        58, 93, 97, 118, 120–121
vision system development
    adaptation to scene and im-
        age properites, 6, 53, 106,
        118
    blackboard system, 127
    bottom-up approach, 53, 60,
        106
    chicken and egg problem, 52
    control structure, 54

data flow, 53
error recovery, 76, 97
expert system, 107
generate-and-test, 53, 106
genetic algorithm, 118
geometric image analysis, 3,
    5, 96, 101, 105, 111, 117,
    124, 127
graceful degradation, 82
initial estimates, 53, 55, 59–
    60, 73–76
interpretation primitive, 127
Markov Random fields, 118
orthographic projection, 22,
    107, 130, 139
perspective projection, 16, 19,
    107, 129
photometric (color) image anal-
    ysis, 5, 51–87, 96, 127
photometric image analysis,
    124
physics-based postprocessor,
    6, 54
physics-based preprocessor, 7
physics-based reasoning, 128
physics-based vision system,
    108, 113, 126–127
segmentation conflict, 62, 95,
    97
top-down approach, 60, 106
traditional system, 106, 127
vision problem, 128
vision programmer's workbench,
    127

T - #0058 - 101024 - C50 - 234/156/9 [11] - CB - 9781568810133 - Gloss Lamination